LIBRARY AND INFORMATION CENTRE

ROYAL SOCIETY OF CHEMISTRY

This book should be returned, or the loan renewed, by the last date shown below. Borrowers must conform with the Library and Information Centre regulations, copies of which can be obtained from the Librarian.

RETURN BOOK BY	RETURN BOOK BY	RETURN BOOK BY

Library and Information Centre, The Royal Society of Chemistry, Burlington House, Piccadilly, London W1J 0BA
Telephone +44 (0)20 7437 8656; Fax +44 (0)20 7287 9798; Email: library@rsc.org

Marine Pollution and Human Health

ISSUES IN ENVIRONMENTAL SCIENCE AND TECHNOLOGY

EDITORS:

R.E. Hester, University of York, UK
R.M. Harrison, University of Birmingham, UK

EDITORIAL ADVISORY BOARD:

P. Crutzen, Max-Planck-Institut für Chemie, Germany, **S.J. de Mora**, Plymouth Marine Laboratory, UK, **G. Eduljee**, SITA, UK, **L. Heathwaite**, Lancaster University, UK, **S. Holgate**, University of Southampton, UK, **P.K. Hopke**, Clarkson University, USA, **Sir John Houghton**, Meteorological Office, UK, **P. Leinster**, Environment Agency, UK, **J. Lester**, Imperial College of Science, Technology and Medicine, UK, **P.S. Liss**, School of Environmental Sciences, University of East Anglia, UK, **D. Mackay**, Trent University, Canada, **A. Proctor**, Food Science Department, University of Arkansas, USA, **D. Taylor**, AstraZeneca plc, UK.

TITLES IN THE SERIES:

1: Mining and its Environmental Impact
2: Waste Incineration and the Environment
3: Waste Treatment and Disposal
4: Volatile Organic Compounds in the Atmosphere
5: Agricultural Chemicals and the Environment
6: Chlorinated Organic Micropollutants
7: Contaminated Land and its Reclamation
8: Air Quality Management
9: Risk Assessment and Risk Management
10: Air Pollution and Health
11: Environmental Impact of Power Generation
12: Endocrine Disrupting Chemicals
13: Chemistry in the Marine Environment
14: Causes and Environmental Implications of Increased UV-B Radiation
15: Food Safety and Food Quality
16: Assessment and Reclamation of Contaminated Land
17: Global Environmental Change
18: Environmental and Health Impact of Solid Waste Management Activities
19: Sustainability and Environmental Impact of Renewable Energy Sources
20: Transport and the Environment
21: Sustainability in Agriculture
22: Chemicals in the Environment: Assessing and Managing Risk
23: Alternatives to Animal Testing
24: Nanotechnology
25: Biodiversity Under Threat
26: Environmental Forensics
27: Electronic Waste Management
28: Air Quality in Urban Environments
29: Carbon Capture
30: Ecosystem Services
31: Sustainable Water
32: Nuclear Power and the Environment
33: Marine Pollution and Human Health

How to obtain future titles on publication:

A subscription is available for this series. This will bring delivery of each new volume immediately on publication and also provide you with online access to each title via the Internet. For further information visit http://www.rsc.org/issues or write to the address below.

For further information please contact:
Sales and Customer Care, Royal Society of Chemistry, Thomas Graham House, Science Park, Milton Road, Cambridge, CB4 0WF, UK
Telephone: +44 (0)1223 432360, Fax: +44 (0)1223 426017, Email: sales@rsc.org

ISSUES IN ENVIRONMENTAL SCIENCE AND TECHNOLOGY

EDITORS: R.E. HESTER AND R.M. HARRISON

33
Marine Pollution and Human Health

RSCPublishing

ISBN: 978-1-84973-240-6
ISSN: 1350-7583

A catalogue record for this book is available from the British Library

© Royal Society of Chemistry 2011

All rights reserved

Apart from fair dealing for the purposes of research for non-commercial purposes or for private study, criticism or review, as permitted under the Copyright, Designs and Patents Act 1988 and the Copyright and Related Rights Regulations 2003, this publication may not be reproduced, stored or transmitted, in any form or by any means, without the prior permission in writing of The Royal Society of Chemistry or the copyright owner, or in the case of reproduction in accordance with the terms of licences issued by the Copyright Licensing Agency in the UK, or in accordance with the terms of the licences issued by the appropriate Reproduction Rights Organization outside the UK. Enquiries concerning reproduction outside the terms stated here should be sent to The Royal Society of Chemistry at the address printed on this page.

The RSC is not responsible for individual opinions expressed in this work.

Published by The Royal Society of Chemistry,
Thomas Graham House, Science Park, Milton Road,
Cambridge CB4 0WF, UK

Registered Charity Number 207890

For further information see our web site at www.rsc.org

Preface

There is growing concern about the state of the world's oceans. The rapid growth of human populations in coastal regions has led to increasing dependence on marine resources, with both positive and negative impacts on human health. Beneficial features related to food supply and life style need to be balanced against the hazards presented, such as microbial pathogens, chemical pollutants from industry and toxic algal blooms. In this book, a group of experts, representing a wide range of disciplines and drawn from several countries in Europe and the USA, present a critical and timely review of the key aspects of the marine environment in relation to human health.

Icarus Allen of the Plymouth Marine Laboratory, UK, presents an overview of the whole of the subject in the first chapter and sets the scene for the more focused and specialised chapters that follow. He points to the many beneficial goods and services to mankind provided by the marine environment but also points to ways in which it poses a risk to the health of coastal populations. This overview demonstrates the need for integration of a wide range of disciplines, from physical oceanography and marine biology to molecular biology and epidemiology, in order to understand and predict the consequences of environmental changes and exploitation of natural resources upon our coastal ecosystems and upon society, including human health.

Jill Stewart of the University of North Carolina, USA, together with several of her colleagues from other prominent American institutions, then give a wide-ranging and detailed review of waterborne pathogens in Chapter 2. The chapter summarises the types of pathogens that are a threat to human health, as well as the fecal-indicator bacteria that are commonly used as surrogates for pathogens in regulatory and research applications. Recent and previous epidemiology studies linking microbial measures of water quality to health outcomes are discussed in detail. The chapter provides an overview of the pathogens, microbial measures and policy implications important for protecting humans from exposure to pathogens in marine waters.

In Chapter 3 James Readman and colleagues from the Plymouth Marine Lab, UK, and the Instituto Español de Oceanografía in Murcia, Spain, discuss the topic of estuarine and marine pollutants. In relation to human health, the consumption of contaminated seafood is the major route of uptake and has implications with respect to the growth of aquaculture. The chapter addresses priority pollutants, emerging contaminants presently under scrutiny (including nanoparticles) and plastics. Climate change implications and their effects on pollution are also investigated and future issues of concern are debated.

Keith Davidson and his colleagues from the Scottish Association for Marine Science in Oban, Scotland, and the Agri-Food and Biosciences Institute in Belfast, Northern Ireland, review the subject of harmful algal blooms (HABs) in Chapter 4. Some species of phytoplankton produce biotoxins which become concentrated in the flesh of organisms (particularly bivalve molluscs) that filter-feed on the plankton. While in most cases there are no adverse effects to these primary consumers, this concentrating mechanism creates a risk to health if the shellfish are consumed by humans. The chapter reviews these issues and the methodologies used to safeguard human health from HAB-generated syndromes.

In the final chapter by Michael Moore and his colleagues from the Plymouth Marine Lab and the Peninsula Medical School, Cornwall, UK, the scientific challenges and policy needs relevant to this area are discussed. Recognising that human health and wellbeing are at risk from increased burdens of bacterial and viral pathogens from sewage and agricultural faecal run-off, as well as chemical and particulate waste from a variety of sources such as industry, domestic effluent, combustion processes, agricultural run-off of pesticides and nutrients, transport and road run-off, the need for new policy formulation to deal with these problems is urgent. The chapter proposes a holistic systems approach, such as Integrated Coastal Zone Management, to address the highly interconnected scientific challenges of increased human population pressure, pollution and over-exploitation of food (and other) resources as drivers of adverse ecological, social and economic impacts, and the urgent and critical requirement for effective public health solutions to be developed through the formulation of politically and environmentally meaningful policies.

We are pleased to acknowledge the guidance given by Steve de Mora, Director of the Plymouth Marine Laboratory, in producing this volume, which we hope will be found useful by all involved in marine science and environmental management and policy.

Ronald E. Hester
Roy M. Harrison

Contents

Editors xi

List of Contributors xiii

Marine Environment and Human Health: An Overview 1
J. Icarus Allen

1	Introduction	1
2	Conceptual Framework	5
3	Issues addressed in this Book	10
	3.1 Pathogens	10
	3.2 Pollutants	11
	3.3 Harmful Algal Blooms (HABs)	12
	3.4 Public Health and Wellbeing	14
	3.5 Scientific Challenges and Policy Needs	15
4	Towards a Systems Approach	16
References		21

Waterborne Pathogens 25
*Jill R. Stewart, Lora E. Fleming, Jay M. Fleisher,
Amir M. Abdelzaher and M. Maille Lyons*

1	Introduction	25
2	Human Pathogens in the Marine Environment	27
	2.1 Pathogens Introduced to the Oceans	27
	2.2 Pathogens Indigenous to the Oceans	30
	2.3 Differentiating Pathogenic from Non-Pathogenic Microbes	30
	2.4 Pathogen Distribution	31
	2.5 Pathogen Detection	32

3	Fecal Indicator Bacteria	32
	3.1 Development and Usage	33
	3.2 Limitations	35
4	Alternative Measures of Microbial Quality	38
	4.1 The Ideal Indicator	38
	4.2 Alternative Indicators	39
	4.3 Microbial Source Tracking	40
5	Molecular Methods: A Revolution in Detection Technologies	42
6	Epidemiological Studies: Linking Microbial Measures to Human Health	44
7	Modeling Pathogens in Marine Waters	53
	7.1 Modeling Aquatic Pathogens: The Example of Vibrios	53
	7.2 Coupling Modeling and Remote Sensing	55
	7.3 Use of Models in Management: Fecal Indicator Bacteria	56
8	The Future of Beach Regulation	56
References		58

Estuarine and Marine Pollutants — 68

James W. Readman, Eniko Kadar, John A. J. Readman and Carlos Guitart

1	Context	68
2	Public Perception	70
3	Priority Substances and Legislation	71
4	Emerging Contaminants	74
	4.1 Industrial Emerging Contaminants	75
	4.2 Other Emerging Contaminants	78
5	Nanoparticles	80
	5.1 Sources and Environmental Behaviour	80
6	Plastics	84
7	Complex Mixtures: Causality of Effects	85
8	Climate Change and Pollutants	86
9	Future Issues	88
References		89

Harmful Algal Blooms — 95

Keith Davidson, Paul Tett and Richard Gowen

1	Phytoplankton	95
	1.1 Harmful Phytoplankton	98
	1.2 Mechanisms of Harm to Human Health	99
	1.3 The Scale of the Problem	99
2	Human Health Syndromes	99
	2.1 Shellfish Poisoning	99

	2.2	Causative Organisms and Toxins	100
		2.2.1 Paralytic Shellfish Poisoning (PSP)	100
		2.2.2 Amnesic Shellfish Poisoning (ASP)	101
		2.2.3 Diarrhetic Shellfish Poisoning (DSP)	102
		2.2.4 Other Lipophilic Shellfish Toxins (LSTs)	103
		2.2.5 Neurotoxic Shellfish Poisoning (NSP)	104
	2.3	Respiratory Illness	105
	2.4	Fish Vectored Illness	105
	2.5	Cyanobacteria	106
	2.6	The Role of Harmful Phytoplankton in Influencing Human Wellbeing	106
		2.6.1 Microflagellates	107
		2.6.2 Other Dinoflagellates	107
		2.6.3 Diatoms	108
3	Harmful Algal Blooms in UK Coastal Waters		108
	3.1	Shellfish Poisoning	108
		3.1.1 Paralytic Shellfish Poisoning (PSP)	109
		3.1.2 Amnesic Shellfish Poisoning (ASP)	110
		3.1.3 Diarrhetic Shellfish Poisoning (DSP)	111
		3.1.4 Azaspiracid Poisoning (AZP)	111
	3.2	Other Harmful Phytoplankton in UK Waters	111
		3.2.1 Karenia mikimotoi	111
		3.2.2 Other Dinoflagellates	112
		3.2.3 Phaeocystis	112
		3.2.4 Microflagellates	112
		3.2.5 Diatoms and Silicoflagellates	113
		3.2.6 Other Species of Pelagic Microplankton	113
4	Safeguarding Health		113
	4.1	Monitoring	113
	4.2	Are Algal Toxins a Public Health Problem?	116
	4.3	Early Warning Methodologies and Mitigation	117
	4.4	Introductions and Transfers of New Species	119
	4.5	Climate Change	120
Acknowledgements			121
References			121

Scientific Challenges and Policy Needs   128
Michael N. Moore, Richard Owen and Michael H. Depledge

1	Introduction	129
2	Key Science Challenges for Marine Environment and Human Health	135
	2.1 Linking Ecosystem Integrity, Ecosystem Services and Human Health	135

	2.2	Sustainable Industrial Development	136
	2.3	Understanding and Mitigating the Impacts of Climate Change	137
	2.4	Better Prediction Systems for Natural Disasters	137
	2.5	Understanding the Distribution and Risks of Marine Biogenic Toxins (Algal Toxins)	138
	2.6	Identifying and Reducing Viral and Bacterial Pathogens from Sewage and Agricultural Run-Off	139
	2.7	Understanding Emerging Risks (e.g. Nanoparticulates from Industrial and Domestic Use)	139
	2.8	Conventional Chemical Inputs (Industrial, Domestic, Agricultural and Road Run-Off), including Personal Care Products, Disinfectants, Pharmaceuticals, Novel Chemicals and Radionuclides	141
	2.9	Endocrine Disruption	143
	2.10	Pharmaceuticals from the Sea	144
	2.11	The Marine Environment as a Health and Wellbeing Resource: the 'Blue Gym' Effect	144
3	Public Health Needs		144
	3.1	Health-Related Indices of Environmental Impact	144
	3.2	Seafood Safety	145
	3.3	Environmental, Social and Economic Interactions (Quality of Governance, Overpopulation and Sustaining Critical Coastal Ecosystems)	145
	3.4	Modelling – Need for an Integrated Approach in the Development of Effective Environmental and Public Health Policies on a Regional and Global Scale	146
4	Policy Needs		146
5	Discussion		150
6	Conclusions and Recommendations		152
References			158

Subject Index   **164**

Editors

Ronald E. Hester, BSc, DSc (London), PhD (Cornell), FRSC, CChem

Ronald E. Hester is now Emeritus Professor of Chemistry in the University of York. He was for short periods a research fellow in Cambridge and an assistant professor at Cornell before being appointed to a lectureship in chemistry in York in 1965. He was a full professor in York from 1983 to 2001. His more than 300 publications are mainly in the area of vibrational spectroscopy, latterly focusing on time-resolved studies of photoreaction intermediates and on biomolecular systems in solution. He is active in environmental chemistry and is a founder member and former chairman of the Environment Group of the Royal Society of Chemistry and editor of 'Industry and the Environment in Perspective' (RSC, 1983) and 'Understanding Our Environment' (RSC, 1986). As a member of the Council of the UK Science and Engineering Research Council and several of its sub-committees, panels and boards, he has been heavily involved in national science policy and administration. He was, from 1991 to 1993, a member of the UK Department of the Environment Advisory Committee on Hazardous Substances and from 1995 to 2000 was a member of the Publications and Information Board of the Royal Society of Chemistry.

Roy M. Harrison, BSc, PhD, DSc (Birmingham), FRSC, CChem, FRMetS, Hon MFPH, Hon FFOM

Roy M. Harrison is Queen Elizabeth II Birmingham Centenary Professor of Environmental Health in the University of Birmingham. He was previously Lecturer in Environmental Sciences at the University of Lancaster and Reader and Director of the Institute of Aerosol Science at the University of Essex. His more than 300 publications are mainly in the field of environmental chemistry, although his current work includes studies of human health impacts of atmospheric pollutants as well as research into the chemistry of pollution phenomena. He is a past Chairman of the Environment Group of the Royal Society of Chemistry for whom he has edited 'Pollution: Causes, Effects and Control' (RSC, 1983; Fourth Edition, 2001) and 'Understanding our Environment: An Introduction to Environmental Chemistry and Pollution' (RSC, Third Edition, 1999). He has a close interest in scientific and policy aspects of air pollution, having been Chairman of the Department of Environment Quality of Urban Air Review Group and the DETR Atmospheric Particles Expert Group. He is currently a member of the DEFRA Air Quality Expert Group, the DEFRA Expert Panel on Air Quality Standards, and the Department of Health Committee on the Medical Effects of Air Pollutants.

List of Contributors

Amir M. Abdelzaher, Department of Civil, Architectural and Environmental Engineering, University of Miami, Coral Gables, FL, USA
J. Icarus Allen, Plymouth Marine Laboratory, Prospect Place, Plymouth PL1 3DH, UK
Keith Davidson, Scottish Association for Marine Science, Scottish Marine Institute, Oban, PA37 1QA, UK
Michael H. Depledge, European Centre for Environment and Health, Peninsula Medical School, Peninsula College of Medicine and Dentistry, Knowledge Spa, Royal Cornwall Hospital, Truro, Cornwall TR1 3HD, UK
Jay M. Fleisher, NOVA Southeastern University, College of Osteopathic Medicine, Masters of Public Health Program, Fort Lauderdale, FL, USA
Lora E. Fleming, European Centre for Environment and Human Health, Peninsula College of Medicine and Dentistry, Cornwall, UK; Oceans and Human Health Center, Rosenstiel School of Marine and Atmospheric Sciences, University of Miami, Miami, FL, USA
Richard Gowen, Agri-Food and Biosciences Institute, Newforge Lane, Belfast, BT9 5PX, UK
Carlos Guitart, Instituto Español de Oceanografía, Centro Oceanográfico de Murcia, c/Varadero 1, Lo Pagan, 30740 San Pedro del Pinatar, Murcia, Spain
Eniko Kadar, Plymouth Marine Laboratory, Prospect Place, Plymouth, PL1 3DH, UK
M. Maille Lyons, Department of Ocean, Earth and Atmospheric Sciences, Old Dominion University, Norfolk, VA, USA
Michael N. Moore, European Centre for Environment and Health, Peninsula Medical School, Peninsula College of Medicine & Dentistry, Knowledge Spa, Royal Cornwall Hospital, Truro, Cornwall, TR1 3HD, UK; Plymouth Marine Laboratory, Prospect Place, The Hoe, Plymouth, PL1 3DH, UK
Richard Owen, European Centre for Environment and Health, Peninsula Medical School, Peninsula College of Medicine & Dentistry, Knowledge Spa, Royal Cornwall Hospital, Truro, Cornwall, TR1 3HD, UK
James W. Readman, Plymouth Marine Laboratory, Prospect Place, Plymouth, PL1 3DH, UK
John A. J. Readman, Plymouth Marine Laboratory, Prospect Place, Plymouth, PL1 3DH, UK

Jill R. Stewart, Gillings School of Global Public Health, University of North Carolina, Chapel Hill, NC, 27599-7431, USA

Paul Tett, Scottish Association for Marine Science, Scottish Marine Institute, Oban, PA37 1QA, UK

Marine Environment and Human Health: An Overview

J. ICARUS ALLEN

ABSTRACT

The marine environment currently provides many beneficial goods and services to mankind but also poses a risk to the health of coastal populations. For example, toxic algal bloom events, microbial pathogens and pollutants all act to negatively impact human health mediated by the marine environment. At the same time, regular contact with the natural environment results in many health benefits, including increased fitness and reduced levels of stress. The marine environment is under pressure from land-derived contaminants and climate change, of which the socio-economic consequences and the implications for human health and wellbeing are not well understood. The scientific challenge is to understand and predict the consequences of environmental changes and exploitation of natural resources upon our coastal ecosystems and upon society, including human health. Addressing this challenge requires the integration of a wide range of disciplines, from physical oceanography and marine biology, to molecular biology and epidemiology.

1 Introduction

Oceans and coastal seas form a vital part of the environment, currently providing many beneficial goods and services to mankind but also posing a risk to coastal populations. They provide a source of natural resources and are a magnet for human habitation. The marine environment's natural grandeur is a source of artistic inspiration, and yet mankind often allows it to be a sink for society's waste. Increasingly, the fragile balance of marine environments is being disrupted by the impacts of climate change and human activities. Interactions between the marine environment and humans are highly significant, diverse and complex. At the same time, the marine environment is strongly affected by the wider global environment (ocean, atmosphere and land) and is sensitive to climate change. Interactions between the atmosphere and the oceans play a significant role in the regulation of climate and weather, which in turn interact with marine environments.

Mankind benefits from the marine environment in many ways. It provides direct tangible benefits, including protein sources and economic activity associated with fisheries and aquaculture. In addition, there are the economic benefits associated with, for example, tourism, renewable energy and recreational activities. Marine ecosystems are a major source of biodiversity and a focal point for biogeochemical cycling. They play a pivotal role in the water cycle and the global biogeochemical cycling of carbon and nitrogen. Other benefits are less obvious and harder to quantify. For example, regular contact with the natural environment results in many benefits, including increased physical activity and therefore fitness, reduced levels of stress, stronger communities and an increased awareness of the value of the natural environment.

Human utilisation of the marine environment also has many negative impacts, with fisheries, industries, agriculture and aquaculture along the world's coastlines contributing to significant physical, chemical and ecological impacts on the surrounding seas. Human activity gives rise to significant inputs of pollutants and pathogens (*e.g.* nanoparticles, radionuclides, bacteria, viruses, nutrients and mixtures of chemical waste), in addition to natural sources (*e.g.* sea birds, marine mammals) to the surrounding seas. Pathogens impact directly on human health through recreational activity: for example, bathing and *via* shellfish into the human food chain. Both of these areas are subject to stringent regulation. Other pollutants and toxins will impact directly on marine organisms, leading to a degradation of ecosystem function and the goods and services provided in terms of carbon and macronutrient cycling, which in turn will impact on fisheries, the recreation value of the marine environment and general human wellbeing. The direct accumulation of pollutants and toxins in human food sources (*e.g.* shellfish) provides pathways for impact on human health. Climate change may exacerbate these effects. Seawater temperature rise may encourage the migration of *Vibrio cholera* and other marine bacteria, such as toxigenic *Vibrios* and *Pseudomonads*, into coastal waters. Increased run-off will also add nutrients to coastal waters, which, together with higher light intensities and temperatures, may increase the growth of toxic algal and cyanobacterial blooms. Flooding associated with sea-level rise and increased storminess may remobilise pollutants from sediments.

The impact/costs of marine environmental degradation are highly dependent on the activities and distribution of the human population and the extent/types of waste water management and environmental regulation. It is therefore important to work towards understanding the pathways and mechanisms through which environmental hazards work and hence be able to communicate these risks to the population at large. Rapid development and unsustainable consumption of natural resources has led to an increasing complexity of environmental health hazards.[1] For example, when an individual person living with poor sanitary conditions in an undeveloped situation consumes seafood contaminated with microbial pathogens, the effects are often apparent. The affected individual will get sick in a short time. In this case the link between cause and effect is relatively easy to measure and, crucially, to communicate. However, if the seafood is contaminated with low levels of persistent organic chemicals which may interfere with human physiology and/or reproduction, the situation is less clear cut. The links between cause and effects are much harder to demonstrate, quantify and communicate. The societal perception of risk presents another great challenge. For example, in the 1950s and 1960s there was a groundswell of public opinion that the UK's bathing waters posed a risk to human health. This was largely attributed to the discharge of untreated sewage into coastal waters. Although the supporting epidemiological evidence for a health risk was not strong, there was an overwhelming belief that swimming in dilute sewage was not healthy.[2]

Climate change is arguably the greatest challenge facing mankind in the twenty first century.[3] In addition to the aforementioned issues, our framework needs to take account of the potential for climate-induced changes in the marine environment.

Resulting from natural variability and anthropogenically induced changes, climate change can only be understood through improved knowledge of the coupling and feedback mechanisms between dynamic processes in the Earth system, as well as the interaction with the anthroposphere. These processes, feedback mechanisms and interactions, in turn, can have unprecedented and dramatic impacts on the marine environment and its ecology and directly on human health.

Marine ecosystems and biodiversity are already under pressure from pollution and overfishing. The marine dimensions of global climate change, such as ocean warming, sea-level rise and changes to ocean chemistry driven in part by atmospheric greenhouse gas concentrations, will influence the marine environment and its impacts on human health. Warmer temperatures and acidification will lead to changes in species reproduction, feeding and, with associated changes in distributions of marine organisms, more frequent algae blooms and shifts in plankton communities. Phytoplankton is a key component of the marine ecosystem, fixing atmospheric carbon and providing the primary food source for the zooplankton, and together they form the base of the oceanic food chain. Larger invertebrates, fish and mammals depend on plankton for their survival. Changing conditions can lead to shifts in the traditional ranges of marine species,[4] resulting in latitudinal shifts in fisheries.[5] In addition, coastal and offshore waters and a range of sensitive marine habitats, such as coral reefs, are likely to be vulnerable to changes in sea-level rise and ocean acidification. Increasing temperature may also influence pollution impacts. For example, it has been shown that warming around the Faroe Islands will facilitate the methylation of mercury, resulting in an estimated 3–5% increase in the mercury content of cod for a 1 °C rise in seawater temperature.[6]

Combinations of such changes will impact on fisheries and aquaculture and will require adaptive measures in order to exploit opportunities and to minimise negative impacts.

The World Health Organisation estimated that over the last 30 years 150 000 lives per year are lost due to anthropogenic climate change impacts on temperature and precipitation. Climatic variations and extreme weather events have profound impacts on infectious disease.[7] The impact of climate change on water quality and quantity is also expected to increase the risk of contamination of public water supplies. Both extreme rainfall and droughts can increase the total microbial loads in freshwater and have implications for disease outbreaks and water quality in estuaries and coastal seas. In particular, infectious agents such as viruses, bacteria and protozoa do not have thermostatic mechanisms, and reproduction and survival rates are strongly affected by fluctuations in temperature. For example, cholera has been shown to vary with climatic fluctuations and sea surface temperatures associated with El Nino Southern Oscillation[8] and many foodborne infectious diseases are sensitive to higher than average temperatures.[7] Finally, coastal tourism will also be affected as a consequence of accelerated coastal erosion and changes in the marine environment and marine water quality, with less fish and more frequent jellyfish and algae blooms.

Although there is clear evidence that climate change will have a significant impact on water quantity and quality, further research is needed in order to ensure that proper decisions on adaptation can be taken. There is a need to improve understanding and modelling of climate changes related to the hydrological cycle and of the water-related impacts of climate change on human health, including their socio-economic dimensions, as well as a need to develop better tools to facilitate integrated appraisals of adaptation and mitigation options across multiple water-dependent sectors. A key issue is that there are large uncertainties in the relationship between atmospheric composition and the resulting climate change. It is highly likely that global climate change will increasingly impact in an all-pervading manner all the connections between the oceans and human health. The diverse and complex issues discussed above require an interdisciplinary and holistic approach to impacts assessments as climatic and societal pressures change in the future.

The scientific challenge is to understand and predict the consequences of environmental changes and exploitation of natural resources upon our coastal ecosystems and upon society, including human health. Addressing this challenge requires the integration of a wide range of disciplines from physical oceanography and marine biology, to molecular biology and epidemiology. Over recent years, this requirement has been increasingly recognised and marine environment and human health is a growing area of interdisciplinary science.[9] By interdisciplinary, I mean the process of answering a question, solving a problem, or addressing a topic that is too broad or complex to be dealt with adequately by a single discipline or profession. This is distinct from a multidisciplinary approach, the act of joining together two or more disciplines without integration. The philosophy should be to engage researchers from multiple disciplines in creating and applying new knowledge, working together as equal stakeholders to address a common challenge. The overall objective of this chapter is to provide an introduction to the current state of research in the marine environment and human health subject area, with an emphasis on interdisciplinary science and the cross-cutting role of modelling and forecasting. Only by integrating a range of scientific disciplines can we hope to understand the present and potential future effects of oceanic and coastal processes and biota on human health and wellbeing, taking account of emerging scientific issues and relevant policy drivers and identifying gaps and synergies in current capability.

2 Conceptual Framework

Management of the marine environment and its consequences for human health requires an understanding of the complex interactions within it. The way in which marine environmental processes and functions interrelate is complex and clearly confounded by their interaction with anthropogenic factors.[10] The potential effects that these may have in terms of human health and wellbeing add a further level of complexity. At the policy level, there is an international consensus, embodied in the EU Marine Strategy Framework Directive, for example, that management of marine resources, in particular management of fisheries, should be conducted on an ecosystem basis. The adoption of an ecosystem-based approach to managing the seas is a step towards addressing these complexities in terms of environmental state. The central tenet of such an approach is the holistic assessment of impacts of human activities on the marine ecosystem and the development of integrated management measures. This requirement is usually translated to mean that, in making management decisions, attributes of the marine ecosystem, such as health, vigour and resilience, should be protected.

As previously stated, the field of the marine environment and human health embraces a wide range of disciplines, from physical oceanography and marine biology to molecular biology and epidemiology, and requires a holistic or 'ecosystem' approach to address it. In order to provide a conceptual framework with which to describe the interaction between the marine environment and human health, I will draw on the Driver–Pressure–State–Impact–Response Model (DPSIR) (see Figure 1). DPSIR was developed for constructing environmental indicators for policy development and implementation of the ecosystem approach to management.[11] Within this model:

- **Drivers:** these describe large-scale, socio-economic conditions and sectoral trends such as population growth, patterns in coastal and watershed land use and land cover, and growth and development in watershed industry sectors.
- **Pressures:** these are the stresses that human activities place on the marine ecosystem, such as anthropogenic climate change, fisheries, aquaculture, wastewater management, the introduction of industrial contaminants and

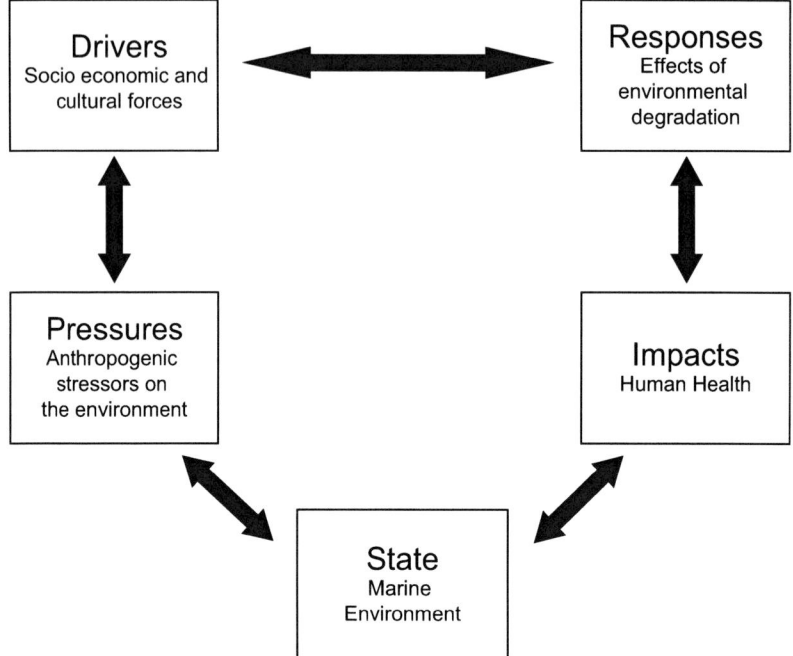

Figure 1 Driver–Pressure–State–Impact–Response (DPSIR) model as applied to the marine environment and human health.

fertiliser use in the coastal watershed which have the ability to directly affect the quality of marine environments.
- **State:** this is essentially the condition of the environment. Indicators of state should describe observable changes in coastal environmental dynamics.
- **Impacts:** these are, for the purpose of this chapter, the effects of environmental state on human health. For example, measured consequences for human health linked to environmental condition such as marine-vectored disease, infections from recreational bathing waters, or exposure to contaminants and toxins *via* fish and shellfish.
- **Response:** this refers to the response of society to the impact of environmental state on human health. It is the institutional response to changes in the system (primarily driven by changes in state and impact indicators).

Coastal and shelf seas are susceptible to a number of large-scale, socio-economic conditions and drivers. It is estimated that over 60% of the world's people live within 100 km of the coast and that this proportion is set to rise as coastal population densities increase.[12] In addition, the world population is expected to increase from about 6 billion to in excess of 8 billion by 2025. Coastal communities are very often heavily reliant on the marine environment for food and raw materials.[13] The dominant driver is population growth, resulting in increased demand for food (fisheries and aquaculture), changes in land use (living space, industry, waste water) and leisure activities (tourism).

The increasing density of coastal populations has and will continue to drive changes in land use. Demand for food and consequent changes in agricultural practice may lead to shifts in fertiliser use and animal husbandry, resulting in alterations to the nutrient supply and the inputs of pathogens to the coastal zone. Increasing populations and associated industrialisation often result in increases in wastewater inputs to coastal seas, resulting in enhanced levels of pollutants and pathogens. It is important to note that the effectiveness of management of wastewater is crucial to the mitigation of the impacts of these effects.

It is estimated that about 400 million people get more than 50% of their animal protein from fish,[14] the large majority of which are caught in shelf/coastal seas.[15] In addition, the sustainability of many remote coastal populations depends on supplies of uncontaminated seafood. World aquaculture has grown significantly during the past two decades, with its growth rate outpacing capture fisheries and other production of animals for food.[14] This rapid growth primarily reflects activities in Asia, in particular in China. The contribution of aquaculture to global supplies of aquatic animals occupied 32% by weight of total fisheries production in 2006. Marine aquaculture accounts for a major portion of global aquaculture production, occupying 55% in 2006 (ref. 16). The downside of marine aquaculture is that it may cause habitat destruction and environmental pollution from its waste products. It is also susceptible to the impacts of natural and man-made contaminants, leading to, for example, algal toxins and pathogens entering the food chain.

Finally, tourism is a major global industry in terms of financial turnover and the number of persons involved. In 2005, the World Tourist Organisation estimated more than 700 million international arrivals, potentially resulting in a greater degree of risk exposure within the global community. Consequently, an increasing number of international travellers potentially now face enhanced risks from consumption of locally harvested seafood and bathing in contaminated waters.[17] A rising population, coupled with a growth in demand for the recreational use of coastal seas, is leading to a rapid increase in the number of people that may be potentially exposed to pathogens, chemical pollutants and algal toxins.

As previously stated, the marine environment is subject to a range of anthropogenic pressures. Anthropogenic pressures, the stresses that human activities place on the marine ecosystem, are a direct consequence of socio-economic drivers. Global change, the result of natural and anthropogenically induced climate change, impacts upon the marine environment and the structure and function of its ecosystems *via* a number of different abiotic and biotic mechanisms. The impacts of some of the key pressures on the marine ecosystems and the subsequent pressures on human health are summarised in Figure 2. These can be sub-divided into climatic and direct anthropogenic pressures. Global change is driven by both natural and anthropogenic increases in atmospheric concentrations of climatically active gases. It will lead to large-scale changes in climate patterns, ocean circulation and climate (*i.e.* structure

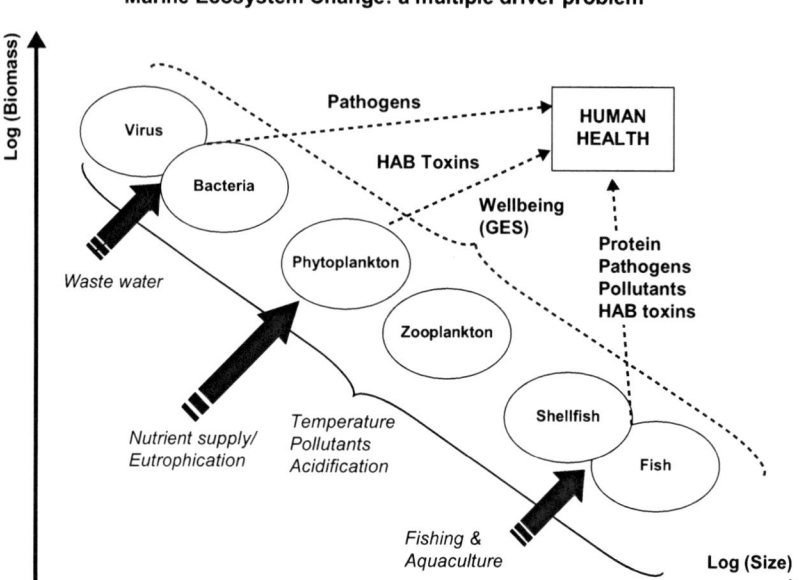

Figure 2 Schematic illustrating where key climatic and anthropogenic pressures impact on the marine ecosystem and the subsequent main vectors impacting human health. The ecosystem is represented in terms of a size spectrum.

and light availability), impacting primary production and propagating change from the base of the marine food web. Enhanced atmospheric CO_2 levels will lead to acidification of the oceans, with significant impacts at all trophic levels from ocean biogeochemistry, calcareous organisms and potentially to the reproductive success of metazoans and fish (*e.g.* changing survival rates of early life history stages). These changes will all impact on biodiversity, the overall trophodynamic structure and functioning of marine ecosystems, and hence on the goods and services they provide to mankind. Simultaneously, combinations of direct anthropogenic drivers such as fisheries, aquaculture, wastewater management, the introduction of industrial contaminants and fertiliser use in the coastal watershed impact at both an organism and population level, thereby influencing the competitive ability and dominance of key species and thus the structure of marine ecosystems. For example, pressures on water quality come mainly from households, industry and agriculture, which use and discharge polluting chemicals and nutrients. For example, high concentrations of sewage or fertilisers in water systems can cause eutrophication, promoting enhanced algal growth, hypoxia and fish death. Waters with high algal biomass turn cloudy and discoloured, losing their aesthetic appeal and recreational function, as well as posing risks to health. One critically important role of the ecosystem is the service it provides in controlling the emergence and spread of infectious disease by maintaining equilibrium between predators and prey, among host vectors and parasites in plants, animals and humans.[18] Biodiversity is essentially acting as a protective function. In the marine system, the spread of cholera has been shown to change in response to increased algal blooms.[19] These effects are secondary, however, when compared with warming seas and changes in eutrophication and sewage discharge. Loss of, or changes in, biodiversity due to anthropogenic pressures may also encourage the proliferation of toxic algae and loss of the aesthetic value of the ecosystem.

There are numerous connections between the marine environment and human activities and mankind which may result in an impact on human health. These impacts may be both positive and negative, resulting in both risks and benefits. The human health impacts of changes in the marine environment can be divided into direct and indirect impacts. Direct impacts include illness derived from poor bathing water quality, pathogens and toxins in the food chain (see Figure 2). Indirect impacts include the health benefits associated with fitness from recreational activities and a general feeling of wellbeing linked to the perceived state of the marine environment. Our current state of knowledge about the linkages between the ocean and human health is highly variable.[20] For examples, the health effects associated with Harmful Algal Bloom (HAB) toxins associated with shellfish poisoning are relatively well understood, as are the social and economic consequences for aquaculture, tourism and fisheries. Less well quantified risks include those linked to chronic exposure to anthropogenic chemicals, pathogens and naturally occurring toxins, where in many cases the mechanisms are poorly understood. This is compounded by a lack of good quality epidemiological data,[20] along with poor information on the social and economic consequences. The extent to which human activities lead to apparent increases in harmful health impacts is subject to considerable debate.

The ultimate challenge is to maintain safe and clean seas which pose no health threat to the human food supply and support healthy and sustainable marine ecosystems. In considering society's response to the diverse range of challenges posed by marine environment and human health, we must take account of the complexity of the issues. The design of management measures to address the risks must consider all relevant drivers, pressures, states and impacts. In responding to marine environment human-health issues, it is of particular importance to understand public behaviour. Public behaviour is frequently governed not by facts but rather by perception. The messages from science are often complex, making it difficult for the issues to be communicated with a clarity that allows firm action to be taken. When managing the risks associated with human health, the central issues are the perception of risk, the response to risk, and the trade-offs between positive and negative effects. The advice given to the public is often contradictory. For example, the health benefits of recreational water sports (*e.g.* swimming, surfing and sailing) on improved fitness are contradicted by warnings of poor bathing water quality and the real or perceived risk of infection from waterborne pathogens. Similarly, general public health messages on the benefits of eating seafood are often contradicted on a local scale, warning populations of the dangers of eating contaminated seafood due to the real or perceived dangers from algal toxins, pathogens and pollutants.

The effective implementation of management responses requires an integrated understanding of the physical, chemical and biological components of the system. This needs to be integrated with an understanding of the behavioural and economic aspects.

Of critical importance to the decision-making process is an understanding of the uncertainties associated with the decision. In many cases, decisions are based on critical thresholds. To get public buy-in and understanding of responses to human-health issues, we need to be able to communicate why these thresholds are critical and identify acceptable levels of uncertainly beyond which integrated impacts decisions cannot be made.

3 Issues addressed in this Book

In the chapters that follow, a number of key issues are discussed in detail: microbial pathogens, pollutants, harmful algal blooms, human health and wellbeing, and scientific challenges and policy needs. By way of introduction, some of the key problems and challenges are outlined in the following sections. The issues and comments raised represent a generalist's viewpoint and are intended to complement the detailed chapters that follow rather than providing a summary of their content.

3.1 Pathogens

The primary health concern with marine microbial pathogens is human exposure associated with both commercial and recreational use of coastal waters.[21] The marine environment provides habitats for a range of microbial pathogens such as bacteria, viruses and parasites. Some of these live free in the water, some attached to particles, while others reside inside larger organisms (*e.g.* shellfish) which are harvested commercially as seafood.[22] Coastal waters contain both natural and anthropogenic sources of pathogens.[23] Natural sources include naturally occurring microbial pathogens, along with faecal pollution from sea birds and sea mammals. Anthropogenic sources include untreated human sewage and agricultural runoff. The numerous pathogens linked to wild, agricultural and domestic animals include a wide range of enteric viruses and waterborne parasitic and bacterial organisms.[23] Contaminated wastewaters come from a variety of point and diffuse sources. While point sources of contamination are relatively easy to indentify and manage, diffuse pollution is more difficult to identify and control. Diffuse pollution is caused predominantly by agricultural runoff and flocks of wild birds.

Most of the known pathogenic viruses that pose a significant public health threat in the marine environment are transmitted *via* the faecal oral route.[22] Agricultural runoff in wet weather can include faeces from grazing cattle and sheep, farm slurries and manures, and sewage sludge put to land. Animal faeces and farm wastes contain large numbers of faecal indicators. Recently, it has been shown that gastrointestinal illness associated with exposure to recreational bathing waters impacted by cattle faeces may pose a similar health risk to those impacted by untreated human faeces.[24] The members of society most susceptible to severe infections are the elderly, the very young and those with compromised immune systems.[21]

The enteric viruses are associated with a variety of human diseases, including ocular and respiratory infections, gastroenteritis, hepatitis, myocarditis and aseptic meningitis. Several species of *Vibrio* are clinically important human pathogens. Most disease-causing strains are associated with gastroenteritis but can also infect open wounds and cause septicemia. They can be carried by numerous sea-living animals, such as crabs or prawns, and have been known to cause fatal infections in humans during exposure. Pathogenic *Vibrio* include *V. cholerae* (the causative agent of cholera), *V. parahaemolyticus*, and *V. vulnificus*. *Vibrio cholerae* is generally transmitted *via* contaminated water. Pathogenic *Vibrio* can cause foodborne infection, usually associated with eating undercooked seafood. Incidences of human viral gastroenteritis have been observed to increase during periods when water temperatures are between 11.5 and 21.5 °C, when shellfish preferentially accumulate microbes.[25] Temperature and light (particularly UVB) are the main bactericides in bathing waters,[2] which explains why the impacts decrease at higher temperatures.

A number of scientific challenges remain in order to understand the impacts of microbial pathogens on human health and these define the future directions of both policy and research. These include indicator organism and their relationships to water quality, non-point sources of contamination, direct pathogen, virulence and non-enteric diseases resulting from recreational water use.[23] The development and application of novel molecular techniques to help elucidate the sources, transport mechanisms and fate of pathogens in coastal waters is helping to improve risk assessments for bathing water quality and seafood consumption.[23]

3.2 Pollutants

A legacy of over 100 years of industrial chemistry is the enormous and diverse range of synthetic chemical compounds which have been released into the marine environment *via* industrial waste, river catchments, sewage and atmospheric sources. Compounds including priority pollutants, POPs (persistent organic pollutants, *e.g.* DDT, PCBs), metals (*e.g.* methylmercury and cadmium), and emerging contaminants such as pharmaceuticals, personal care products, plastics and nanoparticles are major threats to human health. Unlike, for example, pathogens and algal toxins which can cause fatalities, pollution-related human-health issues generally tend to be more subtle. Consequently, potential chemical contaminant effects include longer-term bioaccumulative effects relating to genotoxins and endocrine disruption. The subsequent impacts tend to more chronic (longer term) illness with cumulative/delayed effects. The interactions between humans and pollutants in the marine environment tend to be focused at the point of entry to humans: ingestion, skin contact and inhalation.[26] The most direct pathway of exposure is through the consumption of seafood. For example, the bioaccumulation of PCBs/dioxins and mercury in large pelagic fish potentially exposes the human consumer to cancer risks.[27]

One of the great challenges in assessing the risk of chemical compounds in the environment has been, and remains, the lack of fundamental understanding of the nature, range and scale of pressures imposed on both natural systems and on human health.[28] There is a lack of basic understanding of the spatial and temporal impact of anthropogenic chemicals in marine environments. The increasing proliferation of anthropogenic chemical contaminants into the environment requires a better understanding of how individual compounds and mixtures of pollutants impact on the system. For example, the rapid growth of nanotechnology may present a variety of hazards for environmental and human health.[29] The surface properties and very small size of nanoparticles and nanotubes provide surfaces that may bind and transport toxic chemical pollutants, as well as possibly being toxic in their own right by generating reactive radicals. However, the impact of the release of manufactured nanoparticles into the aquatic environment remains an unknown. Extrapolating from the known behaviours and toxicities for inhaled and ingested nanoparticles in the terrestrial environment, it is likely that there are higher-level consequences for animal health, ecosystems and possible food chain risks for humans.

Similarly, the increasing use of antibiotics in agriculture and aquaculture may result in adverse ecological and human health impacts.[30] Many antibiotics are toxic to aquatic organisms[31] and may impact on ecosystem health and enter the human food chain. While low levels of exposure are unlikely to cause acute toxic effects in humans, the more probable chronic effects remain largely uninvestigated.[31] Finally, the impacts of plastics on marine ecosystem are of increasing concern. Plastics are light, strong and durable and their usage is massive on a global scale. Plastics make up between 60 to 80% of all marine debris;[32] the major concern relating to plastics relates to their ability to absorb and concentrate hydrophobic organic pollutants and act as a carrier for them.[33] As the biological effects of microplastics generally relate to ingestion, they potentially provide another route for contaminants to enter the human food chain *via* consumption of seafood.

To prevent human exposure, a monitoring system is required which provides for the early detection of contaminants.[26] At a global scale, the primary concern is the ingestion of contaminated seafood and the risks associated with it. This requires the rapid assessment of contaminants as early detection is required to prevent human disease. Since analysis of contaminants is expensive and time-consuming, we need to develop appropriate markers and indicators alongside the monitoring and database infrastructure to disseminate and apply such information.

3.3 Harmful Algal Blooms (HABs)

The basic definition of a Harmful Algal Bloom (HAB) is a growth of algae that impacts in ways that humans perceive as harmful. HABs have been associated with fish and shellfish kills, human health impacts and ecosystem damage throughout the world.[34] In recent decades HABs have been expanding, with new dominant species emerging, wider geographic areas being affected, and economic fishery and aquaculture interests seriously impacted. A combination of escalating human activities in coastal ecosystems and the environmental and economic impacts of HABs has exacerbated the challenge for coastal zone management in recent years.

A broad classification of HAB species distinguishes two groups: the toxin producers, which can contaminate seafood or kill fish, and the high biomass producers, which can cause hypoxia or anoxia and indiscriminate mortalities of marine life after reaching high concentrations.[34] Some HAB species have characteristics of both groups. The term 'HAB' is generic and includes species that cause toxic effects even at low cell densities, and recognises that not all HAB species are technically 'algae'. Both types of HABs, the toxin producers and the high biomass producers, have significant impacts on aquaculture, the latter causing massive mortalities of fish and reductions in yields in deteriorated environments, and the former causing shellfish poisonings and potential human-health impacts. Low-oxygen conditions (hypoxia or anoxia) can be produced as the bloom biomass decays and cultured fauna which cannot escape from cages suffer from oxygen stress. Mucilage released by some high biomass producers also causes hypoxia by covering the surface of gills. Suspension-feeding shellfish, including mussels, scallops, oysters and clams, accumulate toxins produced by toxic phytoplankton, sometimes at levels potentially lethal to humans. Toxins and other compounds released by the toxin producers can cause mortality of marine life directly. These damages lead to economic losses and increase operational costs of aquaculture farms due to the expense of monitoring causative organisms, testing to detect algal toxins, and the implementation of countermeasures. HAB toxins in the human food chain can lead to gastrointestinal and neurological effects and, in extreme cases, fatalities. There are four recognised types of shellfish poisoning: paralytic shellfish poisoning (PSP), amnesic shellfish poisoning (ASP), diarrheic shellfish poisoning (DSP) and neurotoxic shellfish poisoning (NSP). All of these are primarily associated with bivalve molluscs. Ciguatera is a foodborne illness, common in the tropical Pacific which is caused by eating reef fishes whose flesh is contaminated with toxins originally produced by dinoflagellates such as *Gambierdiscus toxicus*. These dinoflagellates are eaten by herbivorous fish, leading to the accumulation of toxins up the food chain.

There is evidence that the frequency of HAB events is increasing globally.[35] This is most likely a combination of a genuine increase combined with better monitoring. The reasons behind the increase HAB events are complex. While they are natural phenomena, it is possible that anthropogenic inputs of nutrient (eutrophication) and transport *via* ships' ballast water may exacerbate the magnitude and occurrence of some species.[36] At the same time, climatic and environmental changes may impact on HAB occurrence.[37] Over the last few years progress has been made in determining the effects of environmental conditions on certain HAB species, allowing the development of predictive models of key species (*e.g. Alexandrium*).[38,39] However, the basic physiology and ecology of many species requires further work so we can both construct predictive models and project potential future risk. A second challenge is the continuing development of new methods for the detection of HAB cells and toxins.[40] A final challenge is to understand the human-health effects of chronic sub-acute toxin exposure.[40]

3.4 Public Health and Wellbeing

There are numerous health benefits from eating seafood. Both shellfish and finfish are important sources of proteins, fatty acids and micronutrients (*e.g.* vitamin D, iron, zinc, selenium and iodine). These provide important source of essential nutrients for humans and for many coastal and inland populations, especially in the developing world. While seafood may accumulate contaminants (*e.g.* mercury, ciguatera) and pathogens (*e.g.* norovirus) the overall health benefits to mankind of a well-managed seafood supply chain largely outweigh the risks.[41] While the acute health impacts of human exposure to contaminants in seafood are well understood and often regulated, the potential chronic effects of low-level exposure require further investigation.

The value to humans of leisure in natural settings includes both physical and psychological benefits.[42] Maritime leisure activities (*e.g.* swimming, surfing and coastal walking) are sometimes referred to as the 'blue gym' and growing medical evidence shows that access to the natural environment improves health and wellbeing, prevents disease and helps people recover from illness (see http://www.bluegym.org.uk). Activities involving physical exercise will help to improve cardiovascular health and help to reduce obesity and cancer.[43] Adults who become more active halve their risk of dying early from heart disease. Mental health issues are becoming of significant public health concern worldwide.[44] There is also strong evidence that leisure activities undertaken in a natural environment may help to prevent or improve many mental health issues, for example, reducing stress, improving attention span and psychological aspects such as mood.[27]

In addition, health effects can relate to psychological influences. For example, degradation of natural systems though pollutants and algal toxins are affecting the recreational use of coastal marine waters. In addition, maritime leisure activities (*e.g.* recreational fisheries and scuba diving) can, when poorly managed, adversely affect the marine environment. All of these pressures detract from the human perception of beauty and enjoyment. Loss of amenities, ecology and the potential for reduced economic activity (*e.g.* tourism, aquaculture) can result in human stress, with its associated health implications. Coupled with climate change and extreme weather events, this also affords uncertainty and undermines sustainability.

The challenge for the future is the sustainable management of marine resources in order to protect marine biodiversity and the benefits for human welfare. The implementation of marine protected areas to protect biodiversity and fisheries resources may provide a method of safeguarding the health benefits of healthy seafood and space for recreation for humans.[27]

3.5 Scientific Challenges and Policy Needs

Understanding how human health impacts through, for example, fisheries, HABs and pathogens which are influenced by climate and direct anthropogenic pressures will change in the future, requires a better understanding of ocean and ecosystem dynamics and of their role in accentuating or mitigating climate change. There are two key scientific challenges. The first is the acquisition of basic knowledge and a quantitative understanding of individual processes, pathways, ecological and human-health impacts. The second is the adoption of a systems approach capable of addressing complicated and nested sets of questions. This involves the integration of the wide range of complex processes involved, requiring information and knowledge from the medical, biological, physical, environmental chemistry and the social sciences and is discussed in section 4.

From a policy perspective, the marine environment is increasingly managed through the ecosystem approach coupled with the precautionary principle. The ecosystem approach to management is usually translated to mean that, in making management decisions, attributes of the marine ecosystem, such as health, vigour and resilience, should be protected and this is enshrined in many policy directives, *e.g.* the *Water Quality Framework Directive*, the *Urban Waste Water Directive* and the *EC Marine Strategy Framework Directive* (*MSFD*, 2008/56/EC). For example, the *MSFD* requires member states to develop strategies to achieve a healthy marine environment and make ecosystems more resilient to climate change in all European marine waters by 2020 at the latest. The strategies must contain a detailed assessment of the state of the environment, a definition of 'good environmental status' at regional level, and the establishment of clear environmental targets and monitoring programmes. The precautionary principle implies taking preventative measures when there are reasonable grounds for concern that human activities are leading to risks to environmental status and/or human health.[45] The *MSFD* identifies 11 high-level descriptors (*e.g.* biodiversity, eutrophication, pollution, foodwebs) each of which is characterised by a set of indicators. Such indices may provide potential early indicators of human health hazards.[46] In practice, such ecosystem properties are very difficult to quantify, making operational ecosystem-based management difficult to implement.[47] To make progress, we need a suite of indicators that can be applied serially to detect possible changes to the ecosystem which have implications for human health in response to changes in forcing or to perturbations such as climate change or over-fishing. Ideally such indicators should be objective measures of properties of the system easily understood by both the public and policy makers. The properties of an ideal indicator include representing a well-understood and widely accepted property which is quantifiable unambiguously in standard units, easily measureable with a repeat frequency compatible with intrinsic time-scale of properties under study, and has the potential to create long time series.[48]

In trying to communicate health risk to the wider population, we need to assess how the impacts and uncertainty in the other sectors (climate, water, food, pollution) and the inherent uncertainty in the climate drivers and their variability will impact on integrated health impacts. In particular, attention is required to assess the issues facing decision makers and where they currently derive their information about climate impacts on other sectors that have direct or indirect impacts on health. This requires investigation of the types of information products they require to inform their decision-making, along with an understanding of the critical thresholds and levels of acceptable uncertainty that will need to be achieved to allow decision-making in the health sector across a range of time scales.

Research into the existence, future likelihood and magnitude of the health consequences of anthropogenic pressures and climate change on the natural environment is an important input to international and national policy debates.[49] The recognition of health risks needs to be expanded beyond the economic consequences, loss of amenities and biodiversity to anticipate new threats. Only by doing this will we be able to define adaptive strategies and develop pre-emptive policies.

4 Towards a Systems Approach

As has been previously highlighted, there is a developing awareness of the need to manage the world's seas on a sustainable basis to help ensure the continued diversity and quality of life on Earth. Both natural and man-made stresses may have large functional impacts, resulting in adverse human health and economic consequences. Such effects can arise from exposure to substances that occur widely in marine ecosystems, including biotoxins, endocrine disrupters, heavy metals, nanoparticles, pathogens and synthetic organic pollutants. A wide range of complex processes are involved and their elucidation requires information and knowledge from the biomedical, ecological, physical and social sciences. In order to monitor, understand and predict those impacts, it is necessary to establish the processes and interactions involved. Our core understanding of these issues must be capable of addressing complicated and nested sets of questions, actions and uncertainties.[17] An overarching framework must be capable of embracing questions as diverse as to what and how we eat; human population dynamics; how we determine and measure the quality of the marine environment; how we determine and measure the burden and source of disease; how we collect and manage consumer product information; and how we regulate to mitigate risk.

The development of such a framework requires the addressing of a wide range of integrational challenges. Spatial and temporal challenges range from linking the coastal ocean to both land use and the open ocean, to understanding how molecular-scale processes relate to ecosystem-scale effects at timescales ranging from minutes to centuries. In addition, there are the challenges of linking disciplines, physics to biology, natural sciences to medical and social sciences. These issues then need to be focused towards specific applications. Currently the most commonly used methodology is to take a simple action-reaction approach to a problem.[46] For example, the presence of high concentrations of microbial pathogens which exceed the safe levels stipulated by bathing waters legislation, may lead to the beach being closed until microbial concentrations fall. However, understanding the human health impacts of microbial pathogens is much more complex. Depending on the problem, it may require knowledge of the sources of pathogens; their pathways into the environment; the fate of the pathogens; how they interact with the marine ecosystem; how they come into contact with man; risk of infection; epidemiological studies; and the public perception of risk. This in turn requires a diverse range of knowledge, from tidal dynamics to bioinformatics. Such an approach is necessary if we wish to develop the capability to make short-term predictions of such events, or project future system states under different climate or management scenarios. Finally, there is the challenge of transitioning research to operational applications. These issues need to be placed in the context of both climate change and the rapidly evolving regulatory framework which includes EC directives.

The overall challenge is to develop whole system understanding to enable prediction of changes in key resources which impact human health (*e.g.* ecosystem goods and services, HAB events, pathogen abundance, eutrophication and fish production) under pressure from land-derived contaminants and climate change, the socio-economic consequences and the implications for human health and wellbeing (see Figure 3). The focus of the approach is to enable an integrative rather than reductionist study of the complex interactions that surround natural resource use and hence develop an iterative cycle between prediction and experiment; this is a major science challenge. Crucial to achieving these goals are improved interactions between the environmental and medical research. We identify four key research foci:

- Improved understanding and quantification of sources of pathogens, toxins and pollutants and how they are transported into the marine environment, considering active and passive waste management, river catchments, estuaries and coastal seas.
- Assessment of the impacts of pathogens, toxins and pollutants on environmental status and human food quality. This requires improved understanding of how contaminants are taken up, the pathways both within and between organisms and the impacts on key species and ecosystems.
- Quantification of the consequences of changes in the marine environment (both positive and negative) on human health and wellbeing, taking account of the social and economic impacts.
- A knowledge of how such changes feedback *via* policy and economic activity on the anthropogenic sources, pathogens and pollutants.

The delivery of such a vision requires an environmental systems framework, analogous to the principles of systems biology. Biologists have been reducing life to its constituent parts for over 50 years. The goal of systems biology is to reassemble these data to unravel how complex systems (from sub-cellular processes to organisms) work by developing and evolving series of overlapping

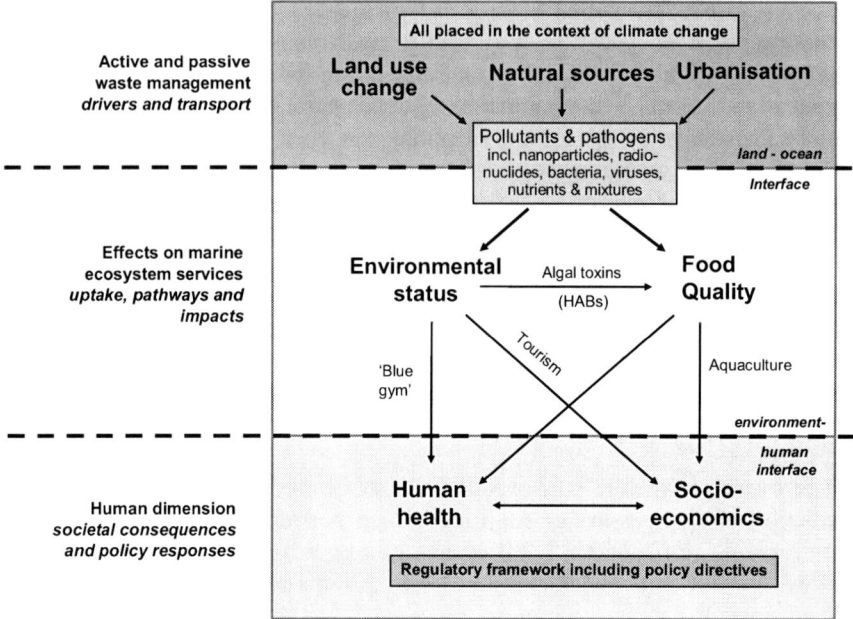

Figure 3 Schematic diagram of a Systems Approach to Marine Environment and Human Health illustrating the interfaces between the land and the ocean, and between the natural environment and medical and social sciences.

conceptual, numerical and statistical models,[50] a process involving the interaction of experiment and simulation in an ongoing iterative process. Environmental Prognostics was proposed by Allen and Moore[51] as a branch of systems biology that is specific to the reactions of organisms to both natural and anthropogenic stress, and the basic principles of which can be extended to human health issues. The key enabling concepts of environmental prognostics[51] are:

1. Acknowledgement that reductionist science acts to disassemble ecotoxicological impacts into constituent processes.
2. Acceptance that biology is a cross-disciplinary science involving mathematics, engineering, information technology and chemistry.
3. Moving towards the notion that biology is an information-based rather than a qualitative science.
4. That the process requires the assembly of systems by modelling followed by disassembly and focused experimentation as an ongoing procedure.

Its role is to provide an integrated explanatory framework for adverse changes in whole systems, from cells to animals to ecosystems, as a branch of systems biology that is specific to the reactions of organisms to both natural and anthropogenic stress. The aim is to develop such frameworks for the evaluation of 'health of the environment' and prediction of consequences resulting from future environmental events and the resultant human-health impacts, based on integrating the reactions of biomarkers for cellular and physiological processes, through conceptual, statistical and computational modelling. These are urgently needed to synthesise complex information on environmental chemistry and injurious effects of pollutants into predicted harmful impact on the health of both sentinel animals and humans.

The crux of the procedure is the definition and evaluation of models of the system in question. This requires the use of the following heavily interdependent tools: conceptual, statistical and numerical models, empirical experimental work and bioinformatics. Bioinformatics is the acquisition, archiving and interpretation of biological information. Statistical analysis is a complementary tool to exploit bioinformatics with the objectives of hypothesis testing, estimation and data exploration. Conceptual models arise from the interpretation of biological data. Numerical models rely on conceptual models to define equations, empirical studies to parameterise them and statistics to evaluate them.

As previously stated, early warning of harmful human health impacts may be obtained by interpreting environmental indicators. For example, the marine ecosystem is sensitive to exposure to toxic contaminants. Pollutants, either individually or in combination, may have sub-lethal effects at the cellular, organ and individual level (*e.g.* causing changes in genetic, behavioural and reproductive activity). Biomarkers include a variety of measures of specific molecular, cellular and physiological responses of key species to contaminant exposure. A response is generally indicative of either contaminant exposure or poor health.[52] However, the lack of effective interpretational capacity has hampered their uptake for use for assessment of risk in environmental management. By integrating individual biomarker responses, we can devise a set of tools and indices capable of detecting and monitoring the degradation in health of a particular organism[53] which, in turn, can be used as indicators of human health risk.

To meet the aspiration of an integrated system, it is clear that we can no longer consider each process and driver in isolation; it is the integrative effects of several drivers which determine the environmental and human-health status. The only tools we have which can integrate different processes and address non-linear combinations of driver impacts in a dynamic environment are numerical simulation models which include dynamic feedbacks. Modelling provides a method by which we can dynamically link sources (*e.g.* pathogens, contaminants) to human exposure. Our knowledge of the impact of the marine environment on human health is currently limited to the climate envelope over which measurements have been made; the use of dynamic simulation models with feedbacks will allow us to assess impacts outside of the observed envelope. In recent years, computational models have played a primary role in the investigation of scientific issues which, given their complexity and the non-linearity of the interactions involved, exceed our native epistemic capacities.[54] Addressing the challenge requires models capable of simulating how non-linear combinations of biotic and abiotic drivers impact on the environment, leading to emergent properties with consequences for human health. Emergence is generally understood to be a process that leads to the appearance of structure not directly described by the defining constraints and instantaneous forces that control a system.

An environmental model is a simplification of a system which describes and simulates its major modes of variability. A good model should describe the available data, represent the system's dynamics and produce emergent properties.[55] A model is a cartoon of a system which hopefully describes the major modes of variability of the simulated system, and careful consideration needs to be made to get an appropriate level of complexity for the questions being asked.[56] Too much complexity leads to uncertainty and problems in interpretation of the model dynamics, but too little means the models cannot reproduce realistic behaviour.[57] To successfully build models of environmental dynamics with a view to both understanding the system and projecting future states, we need suitable data which represent the state of the system in terms of amounts (biomass, concentrations, abundance), rates of change of the system (process rates) and process understanding which links them together. Data and process understanding are the lifeblood of environmental modelling, without which models can neither be parameterised nor have their structure validated or their outputs verified.[56] This requires the systematic collection and storage of data sets, along with the development of methodologies for the intelligent use of such data. To deliver this aspiration, communication between data providers and data users is essential.

A major driver for model development is the demand for quantitative tools to support ecosystem-based management initiatives.[58] Success in ecosystem analyses depends on the ability to integrate and adapt numerous model components in a timely fashion, and the similar basic principles and challenges apply in the context of marine environment and human health. Dynamic models that link the physical, chemical and biological environments provide a means of:

- linking the complex interactions between environmental factors to simulate the growth toxicity and transport of HABs and pathogens;[59]
- understanding how human impacts on different parts of the ecosystem interact (*e.g.* eutrophication, pollution and fishing);
- predicting the consequence of a management action in one sector for other sectors;
- assessing the consequences of alternate management actions; assessing how climate impacts influence progress towards meeting management targets.

While most of the effort in this area is currently applied in terms of general ecosystem response to climate in a fisheries context, there are also examples emerging for pathogens[60,61] and HABs.[62,63]

There is a growing need for real-time information on oceanic and biological processes relevant to human health. The challenge is to provide products which meet the needs of the user which, in turn, brings model predictions and

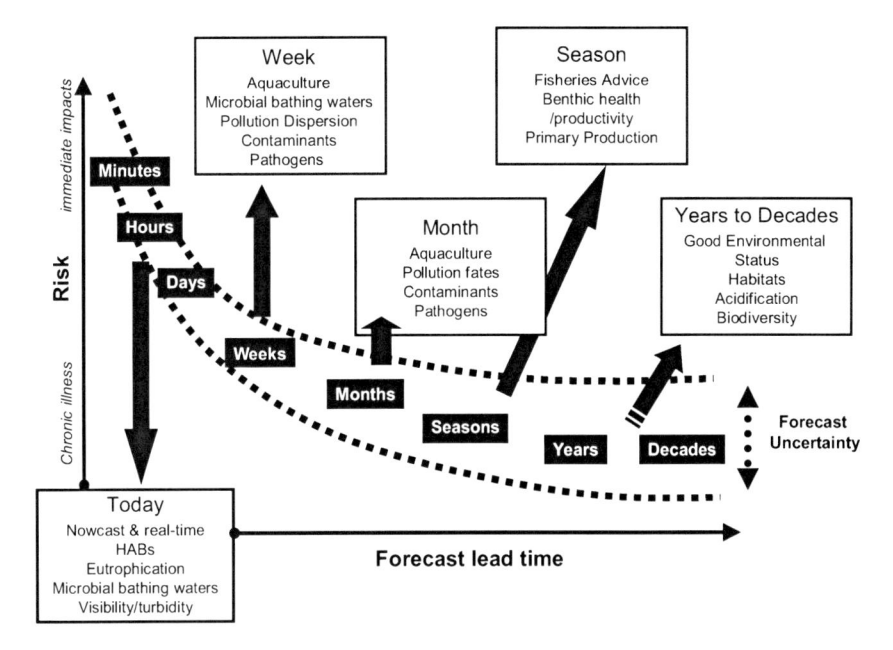

Figure 4 Schematic diagram illustrating the timescales of forecast pertinent to the marine environment and human health.

forecasting, along with real-time data acquisition, to the fore.[59] The range of forecast challenges and timescales is summarised in Figure 4. They range from the short-term forecast and real-time observation of, for example, HABs and microbial pathogens which have immediate short-term human-health impacts, to issues such as fisheries and general environmental status on timescales of seasons to years.

One advantage of an operational approach is that it provides a mechanism to focus research efforts and to translate science into management application, thereby moving the science from being reactive to proactive. As operational services are constructed, there is inevitably a tension between the developer and the user. In terms of operational oceanography, the majority of current systems are strongly developer-led, whereby the focus is on the scientific and technical issues of preparing the forecast rather than the needs of the end-users. In reality, the users often require specific products to be tailored to their needs. In order to do this, we need to identify main user groups/markets and the relevant decision types and timescales. Additionally, we need to encompass other expert knowledge (health, economic, engineering and social) to add value to operational products.

References

1. C. Corvalan, T. Kjellstron and K. Smith, *Epidemiology*, 1999, **10**, 656.
2. K. Jones, *Magazine Health Protect. Agency*, 2008, **12**, 23.

3. IPCC Summary for Policymakers. *Climate Change 2007: The Physical Science Basis. Contribution of Working Group I to the Fourth Assessment Report of the Intergovernmental Panel on Climate Change*, ed. S. Solomon, D. Qin, M. Manning, Z. Chen, M. Marquis, K. B. Averyt, M.Tignor and H. L. Miller, Cambridge University Press, Cambridge, United Kingdom and New York, NY, USA, 2007, http://ipcc-wg1.ucar.edu/wg1/wg1-report.html.
4. G. Beaugrand, P. C. Reid, F. Ibanez, J. A. Lindley and M. Edwards, *Science*, 2002, **296**, 1692.
5. W. W. L. Cheung, C. Close and V. Lam, *Mar. Ecol.: Prog. Ser.*, 2008, **365**, 187.
6. S. Booth and D. Zeller, *Environ. Health Perspect.*, 2005, **113**, 521.
7. J. A. Patz, D. Campbell-Lendrum, T. Hooloway and J. A. Foley, *Nature*, 2005, **438**, 310.
8. M. Pascaul, X. Rodo, S. P. Ellner, R. Colwell and M. J. Bouma, *Science*, 2000, **289**, 1766.
9. E. A. Laws, L. E. Fleming and J. J. Stegeman, *Environ. Health*, 2008, **7**(S2), S7.
10. N. J. Hardman-Mountford, J. I. Allen, M. T. Frost, S. J. Hawkins, M. A. Kendall, N. Mieszkowska, K. Richardson and P. J. Somerfield, *Mar. Pollut. Bull.*, 2005, **50**, 1463.
11. European Environment Agency (EEA), *Global International Water Assessment* 2001, Copenhagen.
12. A. Knap, *Mar. Pollut. Bull.*, 2000, **40**(5), 461.
13. T. Galloway and G. Scally, *Mar. Pollut. Bull.*, 2006, **52**, 989.
14. FAO 2007, *The State of World Fisheries and Aquaculture*, Rome, Italy.
15. R. Watson and D. Pauly, *Nature*, 2001, **414**, 534.
16. GEOHAB 2010, *Global Ecology and Oceanography of Harmful Algal Blooms, Harmful Algal Blooms in Asia*, ed. K. Furuya, P. M. Glibert, M. Zhou and R. Raine, IOC and SCOR, Paris and Newark, DE, USA, 68 pp.
17. R. E. Bowen, H. Halvorson and M. H. Depledge, *Mar. Pollut. Bull.*, 2005, **53**, 541.
18. E. Chivan, *Can. Med. Assoc. J.*, 2001, **164**, 66.
19. R. R. Colwell, *Science*, 1996, **274**, 2025.
20. H. L. Kite-Powell, L. E. Flemming, L. C. Backer, E. M. Fuatman, P. Hoagland, A. Tsuchiya, L. R. Younglove, B. A. Wilcox and R. J. Gast, *Environ. Health*, 2008, **7**(S2), S6.
21. D. W. Griffin, K. A. Donaldson, J. H. Paul and J. B. Rose, *Clin. Microbiol. Rev.*, 2003, **16**, 129–143.
22. A. DePaola, J. Ulaszek, C. A. Kaysner, B. J. Tenge, J. L. Nordstrom, J. Wells, N. Puhr and S. M. Gendel, *Appl. Environ. Microbiol.*, 2003, **69**(7), 3999.
23. J. R. Stewart, R. J. Gast, R. S. Fujioka, H. M. Solo-Gabriel, J. S. Meschle, L. A. Armal-Zettler, E. del Castillo, M. F. Polz, T. K. Collier, M. S. Strom, C. D. Sinigalliano, P. D. R. Moeller and A. F. Holland, *Environ. Health*, 2008, **7**(S2), S3.

24. J. A. Soller, M. E. Schoen, T. Bartrand, J. E. Ravenscroft and N. J. Ashbolt, *Water Res.*, 2010, **44**, 4674–4691.
25. W. Burkhardt, W. D. Watkins and S. R. Ripley, *Appl. Environ. Microbiol.*, 1992, **58**, 826.
26. A. Knap, *Environ. Health Perspect.*, 2002, **110**, 839.
27. J. Lloret, *Mar. Pollut. Bull.*, 2010, **60**, 1640.
28. J. Hyland, L. Balthis, I. Karakassis, P. Magni, A. Petrov, J. Shine, O. Vestergaard and R. Warwick, *Mar. Ecol.: Prog. Ser.*, 2006, **295**, 103.
29. M. N. Moore, *Environ. Int.*, 2006, **32**, 967.
30. A. Sapkota, A. R. Sapkota, M. Kucharski, J. Burke, S. McKenzie, P. Walker and R. Lawrence, *Environ. Int.*, Nov 2009, **34**(8), 1215.
31. O. A. Jones, N. Voulvoulis and J. N. Lester, *Crit. Rev. Toxicol.*, 2004, **34**, 335.
32. J. G. B. Derraik, *Mar. Pollut. Bull.*, 2002, **44**, 842.
33. E. L. Teuten, S. J. Rowland, T. S. Galloway and R. C. Thompson, *Environ. Sci. Technol.*, 2007, **41**, 7759.
34. GEOHAB, 2001, *Global Ecology and Oceanography of Harmful Algal Blooms, Science Plan*, ed. P. M. Glibert and G. Pitcher, SCOR and IOC, Baltimore and Paris, 87 pp.
35. G. Hallegraeff, *Phycologia*, 1993, **32**, 79.
36. D. M. Anderson, P. M. Glibert and J. M. Burkholder, *Estuaries*, 2002, **25**, 562.
37. S. K. Moore, V. L. Trainer, M. J. Mantua, M. S. Parker, E. A. Laws, L. A. Backer and L. E. Fleming, *Environ. Health*, 2008, **7**(S2), S5.
38. D. M. Anderson, D. J. Meafer, D. M. McGillicuddy, K. E. Mickelson, P. S. Keay, J. P. Libby, C. A. Manning, J. M. Whittaker, R. H. Hickey, D. R. Lynch and K. Y. Smith, *Deep Sea Res. II*, 2005, **52**, 2856.
39. Y. Z. Li, R. Y. He, D. J. McGillicuddy, D. M. Anderson and B. A. Keafer, *Continental Shelf. Res.*, 2009, **29**, 2069.
40. D. L. Erdner, J. Dyble, M. L. Parsons, R. C. Stevens, K. A. Hubbard, M. L. Wrabel, S. K. Moore, K. A. Lefebvre, D. M. Anderson, P. Bienfang, R. R. Bidigare, M. S. Parker, P. Moeller, L. E. Brand and V. L. Trainer, *Environ. Health*, 2008, **7**(S2), S2.
41. A. H. Stern, *Environ. Health*, 2007, **6**, 31.
42. World Health Organisation, *The World Health Report*, 2003.
43. J. F. Bell, J. S. Wilson and G. C. Liu, *J. Prevent. Med.*, 2008, **35**, 533.
44. M. Prince, V. Patel, S. Saxena, M. Maj, J. Maselko, M. R. Phillips and A. Rahman, *The Lancet*, 2007, **370**, 859.
45. J. B. Wiener and M. D. Rogers, *J. Risk Res.*, 2002, **5**, 317.
46. M. N. Moore and M. Depledge, this volume.
47. T. Platt, *ICES J. Mar. Sci.*, 2007, **64**, 863.
48. T. Platt and S. Sathyendranath, *Remote Sensing Environ.*, 2008, **112**, 3426.
49. A. J. McMichael, R. E. Woodruff and S. Hales, *The Lancet*, **367**, 859.
50. P. Hunter, *The Scientist*, 2003, **17**, 20.
51. J. I. Allen and M. N. Moore, *Mar. Environ. Res.*, 2004, **58**, 227.

52. M. N. Moore, J. I. Allen and A. McVeigh, *Mar. Environ. Res.*, 2006, **61**, 278.
53. A. Dagnino, J. I. Allen, M. N. Moore, K. Broeg, L. Canesi and A. Viarengo, *Biomarkers*, 2008, **12**(2), 155.
54. J. Symons, *Minds Machines*, 2008, **18**(4), 475.
55. D. Noble, *Biochem. Soc. Trans.*, 2003, **31**, 156.
56. J. I. Allen, J. Aiken, T. R. Anderson, E. Buitenhuis, S. Cornell, R. J. Geider, K. Haines, T. Hirata, J. Holt, C. Le Quere, N. Hardman-Mountford, O. N. Ross, B. Sinha and J. While, *J. Mar. Syst.*, 2010, **81**, 19.
57. E. A. Fulton, D. M. Smith, A. D. M. and C. R. Johnson, *Mar. Ecol.: Prog, Ser.*, 2003, **253**, 1.
58. E. K. Pikitch, C. Santora, E. A. Babcock, A. Bakun, R. Bonfil, D. O. Conover, P. Dayton, P. Doukakis, D. Fluharty, B. Heneman, E. D. Houde, J. Link, P. A. Livingston, M. Mangel, M. K. McAllister, J. Pope and K. J. Sainsbury, *Science*, 2004, **305**, 346.
59. J. Dyble, P. Bienfang, E. Dusek, G. Hitchcosk, F. Holland, E. Laws, J. Lerczak, D. J. McGillicuddy, P. Mennett, S. K. Moore, C. O. Kelly, H. Sole-Gabriele and J. D. Wing, *Environ. Health*, 2008, **S7**(S2), S5.
60. S. B. Grant, J. H. Kim, B. H. Jones, S. A. Jenkins, J. Wasyl and C. Cudaback, *J. Geophy. Res.*, 2005, **110**, C10025–10045.
61. A. B. Boehm, *Environ. Sci. Technol.*, 2003, **37**, 5511.
62. R. He, D. J. McGillicuddy jr, B. A. Keafer and D. M. Anderson, *J. Geophys. Res. Oceans*, 2008, **113**, C07040.
63. J. I. Allen, T. J. Smyth, J. R. Siddorn and M. Holt, *Harmful Algae*, 2008, **8**, 70.

Waterborne Pathogens

JILL R. STEWART,* LORA E. FLEMING, JAY M. FLEISHER,
AMIR M. ABDELZAHER AND M. MAILLE LYONS

ABSTRACT

A variety of microorganisms occur in the marine environment which are capable of infecting humans. This chapter, focused on waterborne pathogens, summarizes the types of pathogens that are a threat to human health, as well as the fecal indicator bacteria that are commonly used as surrogates for pathogens in regulatory and research applications. Limitations and alternatives to traditional fecal indicator bacteria are explored, highlighting challenges and policy implications for protecting public health. Methodological advances and challenges are also reviewed, with an emphasis on research designed to fill gaps and provide scientific support for management of marine resources, particularly bathing beaches. Accordingly, recent and previous epidemiology studies linking microbial measures of water quality to health outcomes are discussed in detail. As an alternative to the measurement of individual water samples, modeling of pathogens in marine waters is introduced. Overall, this chapter provides an overview of the pathogens, microbial measures and policy implications important for protecting humans from exposure to pathogens in marine waters.

1 Introduction

Humans may be exposed to aquatic microbial pathogens (*e.g.* viruses, bacteria, protozoa and fungi) while participating in recreational and occupational activities in the marine environment. Routes of exposure include direct and

---

*Corresponding author

indirect contact with contaminated seawater (*e.g.* skin, eye, respiratory and oral contact) and consumption of contaminated seafood.[1,2] Seafood is implicated in at least 10–19% of the estimated 76 million cases of foodborne illnesses occurring annually in the United States (USA), resulting in approximately 325 000 hospitalizations and 5000 deaths with health-related costs up to $152 billion.[3–9] Although most hospitalizations and deaths have been associated with bacteria (particularly in susceptible subpopulations), viruses account for about half of the illnesses in which a causative agent was identified.[4]

In comparison, swimming and bathing in polluted coastal waters throughout the world results in at least 120 million cases of gastrointestinal (GI) disease and 50 million cases of more severe respiratory diseases annually.[10] In the USA, Dwight *et al.*[11] estimated that the economic burden per GI illness was $36.58, while the burden was $77.76 per acute respiratory disease, $37.86 per ear ailment and $27.31 per eye ailment (2001 values in Orange County, California, USA). The authors also estimated the cumulative public health burden from the excess illnesses associated with coastal microbial water pollution to be $3.3 million per year from just two recreational marine beaches in California. Given *et al.*[12] used epidemiological dose–response models to predict the risk of GI illness at 28 beaches (spanning 160 km of coastline in Los Angeles and Orange Counties, California) and estimated between 627 800 and 1 479 200 excess cases occurred each year. These predictions correspond to an annual economic loss of 21 to 51 million dollars (in year 2000 USD). Recently, Ralston *et al.*[9] have estimated the total annual health costs due to marine waterborne pathogens in the United States at approximately $900 million. Estimates include $650 million due to seafood-borne diseases and $300 million due to GI illness from beach recreation. The authors believe that this estimate should be considered at the lower bound of the true cost due to the significant under-reporting and under-diagnosis of human illness associated with exposure to microbial pollution of marine waters.

This chapter reviews several aspects of current research on human pathogens found in the marine environment, with an emphasis on waterborne pathogens (*i.e.* pathogens that occur in the water column). Firstly, several types of pathogens are described, including naturally occurring agents and those introduced from the terrestrial environment. Then, the use of fecal indicator bacteria (FIB) as regulatory targets and research surrogates is evaluated. The potential for use of alternative microbial indicators is also assessed. Next, the technological advances and limitations of molecular tools are highlighted, since the advent of these methods and technologies means that a broader range of pathogens can be detected and measured than ever before, and faster than ever before. The implications of existing and novel microbial measures are discussed for resource management and beach regulation. Finally, modeling pathogen occurrence (including use of remote sensing technology) is introduced as an alternative to current patchy or intermittent water quality analysis practices. The goal of this chapter is to provide an overview of the pathogens, microbial measures and policy implications of pathogens in marine waters. It is recognized that seafood and sand are also important conduits of human pathogens in the marine environment; however, a detailed discussion of these topics is beyond the scope of this chapter. Readers are referred to recent reviews on seafood[7] and sand[13] for more detailed discussion of these topics.

2 Human Pathogens in the Marine Environment

Pathogens are infectious agents that cause disease (see Figure 1). To date, there are more than 50 aquatic pathogens that have been identified as potential threats to human health, causing ailments such as cholera, pneumonia, GI illnesses, wound infections, sepsis, and skin, ear and respiratory infections (see references in Table 2.1 in Thompson *et al.*)[2,14] Aquatic pathogens are typically classified as either 'introduced' or 'indigenous'. Introduced pathogens are also called 'allochthonous' or 'exogenous' because they are not native to the aquatic ecosystem. In contrast, indigenous pathogens are part of the natural microbial community and are called 'autochthonous'. Both allochthonous and autochthonous organisms have been associated with human disease, but to date most of the documented cases are from allochthonous pathogens introduced from human and other animal sources (see Figure 2).

2.1 Pathogens Introduced to the Oceans

Most described pathogens are introduced into the aquatic environment from the terrestrial environment *via* point and non-point sources (see Figure 3).

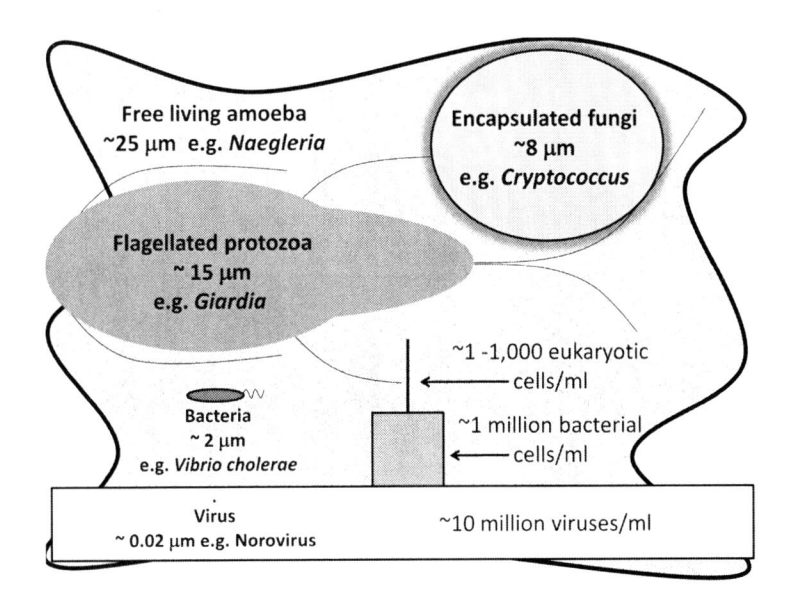

Figure 1 Average numbers (pyramid) of viruses, bacteria and eukaryotic cells expected in 1 milliliter (20 drops) of typical seawater, only a small portion of which are pathogenic and the variety in sizes of representative pathogens.

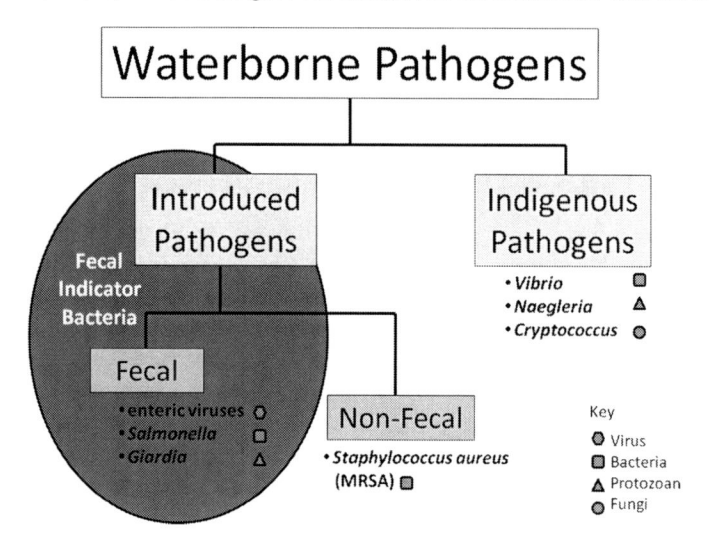

Figure 2 Most of the described waterborne pathogens are introduced and fecal in origin. The use of Fecal Indicator Bacteria (FIB) as indicators for water quality targets only some of the potential risks from waterborne pathogens. Introduced pathogens that are not fecal in origin and indigenous pathogens are not expected to correlate with FIB concentrations. Water quality models for regulatory and management decisions should include all risks.

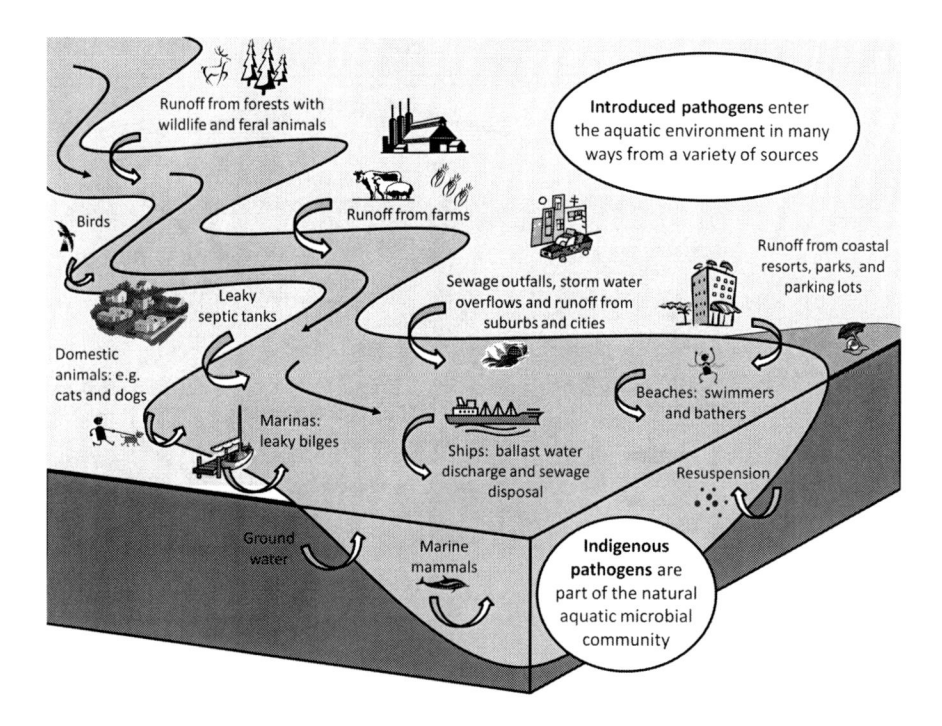

Figure 3 Most human pathogens are introduced into an aquatic environment *via* processes such as runoff, whereas indigenous pathogens are part of the natural microbial community of the ecosystem.

Point sources are specific, identifiable sites, facilities and structures that discharge pathogens and pollutants directly into waterways. Point sources may include sewage outfall pipes, combined sewage-storm water overflow pipes, leaky septic tanks (if identified), drainage outlets and ships discharging sewage or exchanging ballast water. In comparison, non-point sources are diffuse, intermittent and difficult to identify as contributors to the overall pathogen load. Non-point sources may include unidentified leaking septic tanks, unidentified boats, contaminated groundwater discharges, resuspension of sediments and detritus, beachgoers and bathers, and environmental runoff (*i.e.* excess water that flows over land) from forests, farms, suburbs and cities. Assessing the impacts of point and non-point sources on the health of aquatic ecosystems remains a critical challenge for future research.[15] Aquatic pathogens may be further categorized as fecal or non-fecal in origin, and include viruses, bacteria, protozoans and fungi (see Figure 2).

Viruses, ranging in size from 20 to 200 nanometers (0.02 to 0.2 micrometers), are the most abundant biological agents in the marine environment, with typical concentrations of ten billion per liter,[16] but only a small fraction are known to cause disease in humans. Enteric viruses are those viruses that may be found in the GI tract of animals (including humans) and, consequently, shed with their feces. Many enteric viruses are pathogenic; they include the enteroviruses (such as poliovirus, coxsackievirus and echovirus), along with astroviruses, rotaviruses, reoviruses, adenovirus 40/41, hepatitis A and E and the human caliciviruses (such as the Norwalk-like viruses, *i.e.* noroviruses and the Sapporo-like viruses).[17]

Bacteria are prokaryotic cells that are larger (0.2 to 2 micrometers) than viruses, but generally 5 to 25 times less abundant, with concentrations closer to 1 billion per liter (typically 10^6 per milliliter).[16] There are no known pathogenic Archaea, and only a small fraction of Eubacteria cause human diseases. Autochthonous bacterial pathogens include: one or more members of the genera *Escherichia* (*i.e.* pathogenic strains of *E. coli*), *Salmonella*, *Shigella*, *Shewanella*, *Clostridium*, *Campylobacter* and *Helicobacter*.[18–21] Less studied bacterial pathogens that may also pose a public health threat include members of the genera: *Brucella, Burkholderia, Cyclospora, Klebsiella, Listeria, Nocardia, Plesiomonas, Pseudomonas, Streptococcus* and *Yersinia* (Table 17.5 in Santo Domingo and Hansel, 2008; ref. 22). Due to the variety of pathogen types, cost of analysis and difficulty in direct detection, recreational waters are usually monitored for fecal indictor bacteria (FIB), including *E. coli* (all strains) and enterococci, in lieu of any specific pathogen target. The advantages and disadvantages of using FIB to estimate public health risks are addressed below (see section 3).

As eukaryotic cells, the protozoa are generally 10 times larger than bacterial cells, but also less numerous, with typical concentrations ranging from 1000 to 100 000 cells per liter (10^1–10^3 per milliliter).[23] Most research has focused on three problematic enteric protozoa including *Giardia* spp., *Cryptosporidium* spp. and *Toxoplasma gondii*.[24–26] Nearly all disease outbreaks have been associated with drinking- or washing-water contamination, but all three protozoa have been documented in recreational waters. Overall, Giardia spp. are responsible for the majority of cases of identified protozoan-related diarrheal illnesses, both in the USA and worldwide.[27] Other protozoa associated with public health risks include members of the phylum Microspora (including *Encephalitozoon intestinalis, Enterocytozoon bieneusi* and *Vittaforma corneae*) and the amoeboid protist, *Entamoeba histolytica*, the cause of amoebic dysentery.[28]

Examples of non-fecal, introduced pathogens are less common, with the notable exception of the bacterial species *Staphylococcus aureus* which may be shed by bathers at beaches.[29–32] Methicillin-resistant *S. aureus* (MRSA) has also been detected in aquatic environments and, in laboratory experiments, Tolba *et al.*[33] and others have observed survival rates of *Staphylococcus aureus*, including MRSA, to be highest in seawater (compared with that in river and pool water) and to persist for at least two weeks.[34]

2.2 Pathogens Indigenous to the Oceans

Indigenous pathogens are part of the natural microbial community of the local ecosystem and, as a result, eradication of these microbes is not a sensible or feasible goal. Furthermore, concentrations of indigenous pathogens are not expected to correlate with the concentrations of FIB, posing an unquantifiable risk to human health unless specifically targeted during monitoring efforts. Despite their numerical dominance, there are no known indigenous viruses that cause disease in humans.[35] Examples of autochthonous bacterial pathogens include: one or more members of the genera *Aeromonas, Mycobacterium, Legionella, Pseudomonas* and *Vibrio*, including *V. cholerae* (the etiological agent of cholera), *V. vulnificus* and *V. parahaemolyticus* (often responsible for GI illnesses associated with the consumption of contaminated shellfish).[36–41] Another naturally occurring bacterial species that may cause diseases in humans is *Francisella tularensis*, the etiological agent of tularemia (*i.e.* rabbit fever).[42]

Pathogenic indigenous protozoa include free-living amoebae from the genera *Acanthamoeba, Balamuthia, Hartmannella, Sappinia* and *Naegleria*.[43,44] In addition to being human pathogens, some of these protozoa may carry viruses[45] and pathogenic bacteria, such as *Legionella* spp., *Francisella, Mycobacterium, Burkholderia, Vibrio, Listeria, Helicobacter* and pathogenic *E. coli*.[46,47]

Far less is known about marine fungal pathogens, but there have been reported cases of skin infections from *Candida albicans, Malassezia* (formerly known as *Pityrosporum) furfur* and *Microsporum canis* following exposure to recreational waters.[48] More recently, basidiomycete fungal pathogens, *Cryptococcus neoformans* and *Cryptococcus gattii*, which may cause pulmonary and nervous system diseases in humans and animals, have been detected in the environment, including samples from creeks, rivers, lakes and seawater.[49]

2.3 Differentiating Pathogenic from Non-Pathogenic Microbes

The presence of an agent listed above does not necessarily indicate a threat to human health because not all strains of each species are equally virulent.

Furthermore, some human and animal hosts are more susceptible to certain pathogens than others; for example, children, persons with suppressed immune systems (*e.g.* AIDS) and other vulnerabilities (*e.g.* chronic liver disease) may be particularly susceptible to certain pathogens.[50] In general, for a potential pathogen to cause disease in a host, it must: (1) encounter the host; (2) gain entry into the host; (3) spread from the site of entry; (4) multiply, either before or after spreading; and (5) damage the host, either directly or indirectly *via* triggering a host response.[51] Virulence factors are properties or strategies of a pathogen that facilitate these actions, including: capsules, cell surface proteins and carbohydrates that aid in attachment or entry; endo- and exo-toxins; hydrolytic enzymes; hemolysins; motility; and quorum sensing. One way to distinguish pathogenic and non-pathogenic strains is to test for the presence of a virulence factor. For example, there are over 200 strains of *Vibrio cholerae*, but only two serotypes (O1 and O139), distinguished by the structure of the O antigen (*i.e.* carbohydrate in the cell wall), cause epidemics of cholera.[52]

Another way to distinguish pathogenic and non-pathogenic microbes is to monitor for specific genes that code for the production of a virulence factor. For example, to differentiate between total and pathogenic *Vibrio parahaemolyticus*, samples are screened for the gene which produces a virulence factor known as a thermostable direct hemolysin (denoted *tdh*+ if present).[53] Other targets have been developed for pathogenic strains of *Aeromonas hydrophila* (aerolysin gene: *aero*), *Shigella flexneri* (invasion plasmid antigen H gene: *ipa*H), *Yersinia enterocolitica* (attachment invasion locus gene: *ail*) and *Salmonella typhimurium* (invasion plasmid antigen B gene: *ipa*B) bacteria.[54] For bacteria in general, the presence of virulence factor-related genes has been shown to be associated with genomic islands (*i.e.* groupings of genes) for genera that colonize multiple habitats (*e.g. Vibrio*, *Pseudomonas* and *Campylobacter*), but not for obligate intracellular pathogens (*e.g. Mycobacterium* and *Legionella*), supporting the importance of horizontal gene transfer in the environment.[55]

2.4 Pathogen Distribution

Local and global environmental conditions (such as rainfall, temperature, salinity, nutrient concentrations, solar radiation, pH, wind, tides, currents and biogeography) impact the type, timing, amount and survival of pathogens in aquatic ecosystems. An understanding of how these environmental factors affect the abundance, activity and persistence of pathogenic microbes is vital to predicting their fate.[56] Research targeting pathogen ecology has focused on the prevalence and persistence of the pathogen, because the longer a pathogen survives in the environment, the more likely it is to come in contact with a new host. Pathogenic microbes may be found free-living in the water; within coastal sediments; attached to small particles floating in the water; associated with larger detrital-based organic aggregates suspended in the water; associated with living phytoplankton and zooplankton; inside biofilms on the surfaces of other organisms, rocks or structures; and within other organisms.[35,57-61]

2.5 Pathogen Detection

Proper assessment of human health risks from exposure to the marine environment requires routine evaluation of the sanitary quality of water. However, it is not currently practical to directly identify and enumerate the hundreds of pathogenic microorganisms that may be present. Furthermore, technical difficulties, analysis costs and the need for extensive training and retraining of the labor force are significance hindrances in routinely monitoring for pathogens.

3 Fecal Indicator Bacteria

Fecal indicator bacteria (FIB) have been used for over a century as surrogates for enteric pathogens for both regulatory and research purposes. These non-pathogenic microbes are found in large numbers in the feces of all warm-blooded animals and hence also in human sewage. Some of the estimated levels of FIB in raw human sewage are presented in Table 1. The presence of FIB has been used to indicate the presence of human and animal fecal pollution, therefore suggesting the concurrent presence of enteric pathogens and the risk of exposure and disease in humans.

Fecal coliform and fecal streptococci (including enterococcus) are the two most common groups of FIB used in testing the extent of contamination of water bodies. These organisms may be isolated and quantified using simple culture-based microbiological methods that require minimal training and take 24 to 48 hours to complete (see Figure 4). Fecal coliforms, which include *Escherichia coli*, are differentiated in the laboratory by their ability to ferment lactose with the production of acid and gas at 44.5 °C within 24 hours.[62] Fecal

Table 1 Estimated levels of fecal indicator bacteria (FIB) in raw sewage (Maier *et al.*)[177].

Fecal Indicator Bacteria	Colony forming units (CFU) per 100 ml
Coliforms	10^7-10^9
Fecal Coliforms	10^6-10^7
Fecal Streptococci	10^5-10^6
Enterococci	10^4-10^5
Clostridium perfringens	10^4
Staphylococcus (coagulase positive)	10^3
Pseudomonas aeruginosa	10^5
Acid-fast bacteria	10^2
Bacteroides	10^7-10^{10}

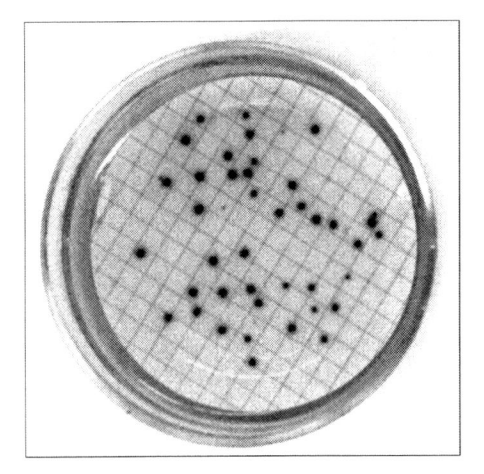

Figure 4 The fecal indicator bacteria (enterococci) growing in colonies on enterococci-specific media (mEI agar). (Courtesy of Matthew Phillips).

streptococci are differentiated by their ability to grow in 6.5% sodium chloride, at a pH of 9.6 and at 45 °C and include: *Ent. avium*, *Ent. faecium*, *Ent. durans*, *Ent. facculis* and *Ent. Gallinarium*.[62] Fecal streptococci are able to persist in harsher environments and at higher salinities than fecal coliforms and have been found to correlate with adverse health effects amongst bathers in marine waters with known point sources.[63] Although fecal coliforms were widely used in the past to monitor marine waters, fecal streptococci, specifically enterococci, are currently the recommended and most widely recognized FIB for monitoring marine waters.[64–66]

3.1 Development and Usage

Epidemiological studies have been conducted in order to determine the levels of FIB in marine bathing waters which would cause adverse health effects above a certain 'acceptable' risk level, as discussed in section 5. Guidelines and criteria were then formalized by national and local government agencies and through the passing of legislation, such as the Clean Water Act and BEACH Act in the USA[67] and the EU directive in the European Union.[66] In short, multiple studies have shown that indicator microbes, particularly enterococci, can demonstrate a significant association with illness at point-source beaches in temperate climates.[68–72] As a result, using enterococci as an indicator for human health impacts (and therefore as a surrogate for pathogens) may be appropriate when recreational beaches are dominated by point sources of sewage contamination (*e.g.* outfalls from sanitary sewer facilities) and in temperate climates.[68–72]

The current standard for enterococci was originally established through a USA epidemiological study in 1982 (ref. 63), which showed a correlation between the indicator bacteria, enterococci, and adverse health impacts in humans at several point-source marine beaches located in a temperate climate. Based on this one study, the United States Environmental Protection Agency (US EPA) adopted a geometric mean standard of 35 colony forming units (CFU) of enterococci per 100 ml of swimming water, and a single sample standard of 104 CFU per 100 ml for marine beaches. The geometric mean value of 35 CFU per 100 ml was equivalent to a risk of 19 illnesses per 1000 people, which was deemed acceptable by the US EPA.[50,73–75] The single sample maximum is based on a percent log standard deviation of the threshold value, and means that no single sample should exceed an Enterococci concentration of 104 CFU per 100 ml, regardless of the geometric mean. Many states in the USA adopted this standard and currently monitor for enterococci as part of the 'Healthy Beaches' programs (http://water.epa.gov/type/oceb/beaches/beaches_index.cfm).

In the United States, approximately 3000 beaches are tested at least weekly for FIB as part of monitoring programs (including the US EPA/State-supported 'Healthy Beaches Program'), resulting in some 18 500 days of closures or advisories in 2009 being issued for recreational beaches.[76] Figure 5 displays the number of beaches monitored, closed, or posted for advisories in the USA during the past 20 years.[67,76]

As an example of the usage of FIB in marine beaches and their ability to predict the presence of human fecal pollution, a study conducted in Miami, Florida, in 2001 may be taken as a case study.[77] This study was conducted to determine levels of FIB at a marine beach over a period of two months. However, during this study, an estimated 8 million gallon sewage spill occurred over a period of two hours near the beaches that were being studied, due to the

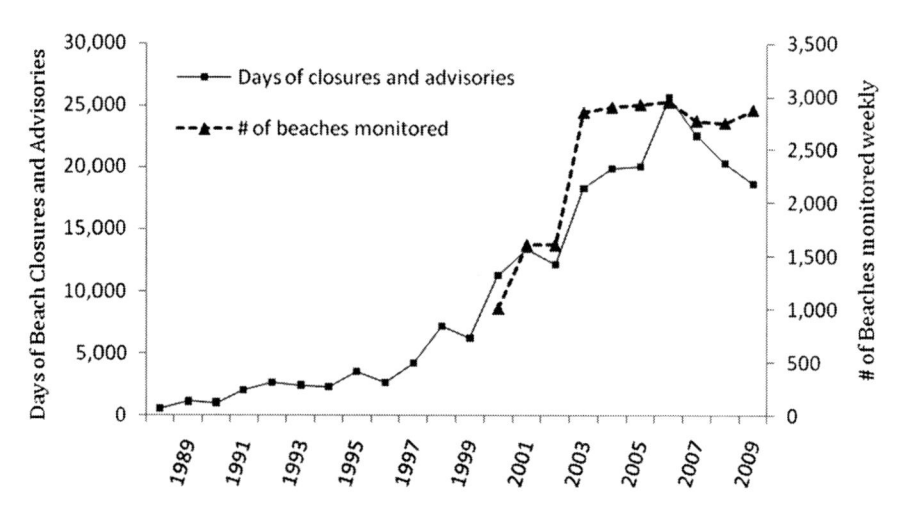

Figure 5 Number of beaches monitored, closed, or posted for advisories in the US (based on Boehm et al.[67] and NRDC).[76]

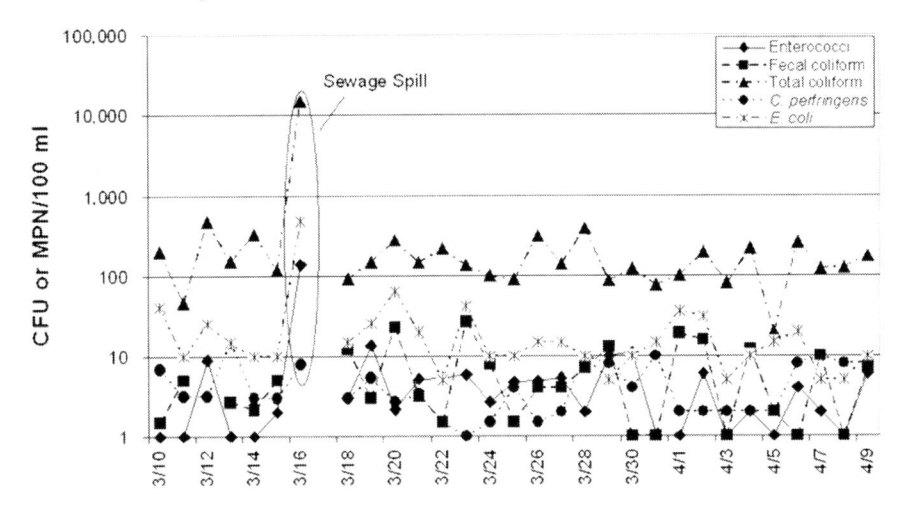

Figure 6 FIB levels during a study at a Florida beach, USA, in which a sewage spill occurred near the study site. (Reproduced from Shibata et al.).[77]

rupture of a 1.8 m diameter force main. The data collected during the two days after the sewage spill event showed significant spikes of several FIB (including total coliforms, *E. coli* and enterococci). Enterococci, used by regulators at this beach for monitoring, exceeded the monitoring criteria (104 CFU per 100 ml) after the sewage spill (see Figure 6). Therefore, in this case study with a known point source, FIB would have been able to successfully predict the presence of human fecal pollution at a marine beach in the extreme scenario of an actual sewage spill.

3.2 Limitations

The concept upon which FIB usage is based is illustrated in Figure 7. Both infected and non-infected individuals release FIB into the sewage stream, but only those who are infected also release enteric pathogens. Consequently, sewage usually has high levels of FIB and substantially lower and more variable levels of specific pathogens. In addition, pathogens present are a function of the illnesses affecting a community at any given time, compared to FIB which are present in high numbers regardless of illness rates. Failure of, or leaks in, sewage treatment facilities or sewage piping systems may result in the input of untreated human sewage into marine water bodies. People coming into contact with water contaminated with human sewage may then become infected through the ingestion, skin contact and/or even inhalation of pathogens. Moreover, even treated sewage still contains both indicator and pathogenic microbes, although at significantly lower levels when compared to untreated sewage.[62] Elevated concentrations of FIB in a water body are used to indicate the possible presence of enteric pathogens. Therefore, regulators

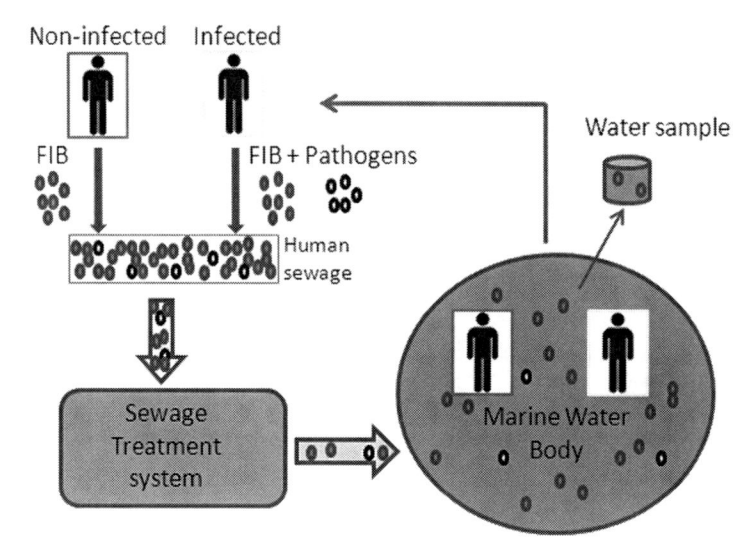

Figure 7 Cycle illustrating the release of FIB and pathogens from individuals which are then transported through the sewage treatment system (sewage lines, septic tank, wastewater treatment plant). Microbes which are not deactivated through the treatment process or those being transported through leaks will enter the marine water body. Pathogens which enter the marine water body may cause infection in bathers. Water samples are taken by regulatory agencies which test the level of FIB in the water and, based on the concentration, decide whether or not to close or place warning signs at the water body.

traditionally use FIB to allow or prevent public access to the water body (or harvesting of potentially contaminated shellfish), or at least warn of the potential risk of coming into contact with the water (see Figure 8).

Although FIB have been instrumental in identifying human fecal pollution in marine waters and therefore a means to protect public health, several limitations and unanswered questions remain regarding their usage. Some of the most fundamental limitations relate to the link between concentrations of FIB and the effect on human health. FIB standards, which are based on threshold concentrations above which the health risk from waterborne illness is unacceptably high, were developed through epidemiological studies to identify risk of developing GI illness.[63,78,79] However, the relationship between FIB concentrations and rates of GI illnesses may not be applicable to all marine beaches of varying climates and pollution sources (*e.g.* sub-tropical or tropical marine environments and non-point sources of pollution such as run-off).[67,72,80] This issue is discussed in more detail in section 5.

Another limitation of FIB is that they are found in all warm-blooded animals, and therefore may enter marine water bodies from sources other than human fecal pollution.[81] These other sources of FIB may or may not contain human pathogens, and therefore elevated concentrations of FIB may not

Figure 8 Warning signs in three different languages due to exceedances of FIB standards at a Florida beach, USA. (Courtesy of Helena Solo-Gabriele).

always be associated with an increase in human health risk. It has also been shown that FIB are heavily influenced by environmental factors such as tides, rain and solar radiation.[82] These factors affect the distribution of FIB and complicate their ability to predict the presence of human fecal pollution and associated illnesses. Studies evaluating the variability of FIB at marine beaches have also demonstrated that FIB levels can vary significantly temporally, within a few minutes,[83] and spatially, between knee- and waist-deep samples.[84] Therefore, depending on when and where samples are taken, conclusions about the quality of the water body can vary greatly. Furthermore, in both sub-tropical and temperate climates, indicator bacteria can multiply in the environment (including in the beach sand in the swash zone), giving a false impression of an increase in fecal pollution.[85–88] This issue is especially problematic since many pathogens of concern, which require human hosts, may not be able to multiply in the environment in a similar fashion.

Therefore, increases in the concentrations of FIB, which are not necessarily associated with increased levels of pathogens and/or human health risks, can lead to unnecessary economic losses at recreational beaches. Beach closures in general can be very expensive in terms of resources and public trust. For example, a four-month closure of Huntington Beach in 1999 resulted in the loss of millions of dollars in tourism income to the business community, and almost two million dollars in beach closure investigation fees.[89,90]

On the other hand, the lack of FIB can be misinterpreted as the absence of a human fecal pollution source, and the absence of pathogenic microorganisms causing human health risk. This may pose a problem given that FIB are less resistant than some pathogens (*e.g. Cryptosporidium spp.* and enteric viruses) to wastewater disinfection at water treatment plants and to environmental stresses.[91,92] Failing to properly close the beach when pathogens are present may result in a significant health risk to bathers and other beach users, and millions of dollars in medical costs and subsequent attendant business and tourism impacts.

An additional concern regarding the usage of FIB is the lack of a causative link between FIB and human illness, since FIB themselves are non-pathogenic, and therefore are not directly the cause of disease. The use of FIB to predict human fecal pollution exposures and human illness would be more justifiable if a consistent association had been demonstrated between FIB and pathogens in marine waters. Research has shown, however, that indicators are not always correlated with pathogens of concern,[72,80,93–95] especially in areas impacted by non-point sources of indicator organisms (*e.g.* rainfall runoff, animals, human shedding, beach sand release, *etc.*).

Limitations also exist with regard to the analysis of FIB. Traditional culture-based methods, which are most widely used for FIB analysis, take 24 to 48 hours to yield results. This would mean that bathers would potentially be in contact with contaminated water for up to 24 to 48 hours before it were known if the water was contaminated. Research is currently being conducted to develop faster molecular-based methods to analyze for FIB and to yield results within a few hours, allowing public notification of exposure and health risk from microbial pollution within a more reasonable time-frame of the health risk associated with the marine water body (see section 6).

4 Alternative Measures of Microbial Quality

4.1 The Ideal Indicator

Given the limitations of using traditional FIB, several alternative indicators have been proposed to amend or replace current procedures. However, in proposing alternative indicator microbes, it is important to understand some of the characteristics which an ideal indicator should possess in order to properly serve as an indicator of pathogens, fecal pollution or human health risk. Most importantly, the concentration of the indicator should be associated with increased adverse health effects in humans as demonstrated through epidemiological studies (discussed in section 5). Additionally, the indicator should be consistently present in all human fecal pollution, as well as fecal matter of other warm-blooded animals which have been known to carry human pathogens. The indicator should also be detected in an amount which is proportional to the level of fecal pollution to allow for quantification of risk. The ideal indicator should have similar resistance to wastewater treatment processes, as well as environmental pressures such as temperature, salinity and solar radiation, when compared to pathogens of concern. The ideal indicator should also lack the ability to regrow in the environment, as this may yield a false impression of increased fecal pollution, and should always be present in the presence of enteric pathogens. Finally, this indicator should be simple to detect in a minimal time frame (within hours or in real time), and at a minimal cost to facilitate routine monitoring.[62,69–71]

4.2 Alternative Indicators

Several classes of alternative indicators have been proposed with varying advantages, disadvantages and unanswered questions. Alternative indicators with the highest potential for success are described below.

Clostridium perfringens is a spore-forming anaerobic bacterium found in human and animal fecal pollution that has been shown to not grow in the marine environment. As a result, the Hawaii State Department uses *C. perfringens* to monitor their water bodies for the presence of fecal contamination.[96] The observation that *C. perfringens* spores are environmentally resistant makes it a suitable indicator for environmental- and treatment-resistant pathogens such as *Cryptosporidium* oocysts (resistance to chlorine) and adenoviruses (resistant to UV radiation).[50] However, this lengthy persistence is also a disadvantage because it cannot be used to differentiate between historic and recent fecal pollution.

Bacteroides spp. are anaerobic bacteria which are among the most prevalent genera found in feces of warm-blooded animals,[97] and do not survive for long periods of time under aerobic conditions.[98] The advantage of using *Bacteroides* spp. as an indicator is that they can be used to differentiate between fecal sources, since there are species that are specific to humans, birds and dogs (as discussed in the following section). However, the detection of *Bacteroides* spp. using qPCR methods (which detect both living and dead microorganisms) in ambient water could also represent historic pollution.

Coliphages are a type of bacteriophage (*i.e.* bacteria-specific viruses) that infect *E. coli* and other similar coliform bacteria.[50] These viruses are present in high concentrations in sewage, persist longer than traditional FIB, and may be easily detected from small sample volumes. Another advantage is that they can be detected by simple and inexpensive culture methods similar to techniques used to detect traditional FIB. Coliphages have been recommended as indicators of enteric viruses because their structure, morphology and size resemble that of enteric viruses.[62] Coliphages that infect *E. coli* via the cell wall are called 'somatic coliphages,' while those infecting the appendages (F-pili) of *E. coli* are called 'male-specific' or 'F+ coliphages.' Recent epidemiological studies at non-point-source polluted beaches have shown weak associations between F+ (ref. 71) and somatic[99] coliphages with reports of GI illness in bathers. However, the ecology of both somatic and F+ coliphages, including their association with fecal contamination and pathogens, is inadequately understood.[50]

Another virus which has been proposed as an indicator is human polyomavirus.[100] These are viruses which are found in human urine and are widespread in human populations. Polyomavirus is the sole genus in the Polyomaviridae family of viruses. Polyomavirus has been detected in sewage as well as polluted marine water samples.[101,102] Finally, because this virus directly infects humans and not human bacteria (*e.g.* coliphages), the presence of polyomaviruses may be more indicative of a human source of fecal pollution. Other potential viruses proposed as indicators of sewage include pepper mild mottle virus (a plant virus common in the human gut),[103] and human enteric viruses, such as enterovirus,[104] adenovirus,[105,106] and norovirus.[104]

Other non-biological indicators of human fecal pollution include several types of chemical markers, such as caffeine and optical brighteners. Although most caffeine is metabolized when consumed, approximately 10% of ingested caffeine remains intact when excreted into the sewage system.[50,107] Optical brighteners are whitening agents added to detergent to offset the yellowing of clothes. Effluent from washing machines is mixed with that of toilets before being transported to wastewater treatment plants, so optical brighteners may be a useful tracer of sewage. Analysis of these markers could provide more timely results than microbiological markers which require time to grow before analysis can be completed. The biggest disadvantage of using chemical indicators is that wastewater treatment plants are generally designed for the removal of microbes to decrease biological oxygen demand (BOD). The plants were not designed for the complete removal of chemicals and the resulting chemical signatures may not reflect the removal of pathogens. Other disadvantages include the dilution factor when introduced to environmental waters and the potential for natural, environmental compounds to interfere with signals from the chemical indicators.[108]

4.3 Microbial Source Tracking

Detection of fecal indicators or pathogens along the coast does not always provide information about the source of pollution. This is particularly true in areas where there is an absence of an obvious point source of pollution or sewage spill. However, information about sources is necessary for managers to effectively address pollution problems.[109,110] Some indication of source is necessary for managers to implement appropriate mitigation measures in order to bring contaminated waters into compliance with regulatory standards. Identification of pollution sources is also useful for the characterization of human health risks. Although a number of pathogens are known to be associated with animal wastes, currently (and perhaps mistakenly, given the high levels of antibiotic-resistant organisms in animal wastes) human-source waste is generally considered to pose a greater risk to human health.[111] Identification of a low amount of human waste may prompt a different management response than the identification of a low amount of animal wastes.

Microbial source tracking is used to determine sources of fecal contamination in the environment, including coastal waters and shellfish. This concept is not new, with attempts dating back to the 1960s when researchers proposed using ratios of fecal coliforms to fecal streptococci to discriminate human from animal fecal contamination.[112] This early approach was limited by the heterogeneity and variable survival rates of the bacterial species involved. However, recent advances in molecular biology have enabled more sensitive typing methods to evolve.[113] Current methods generally involve analysis for a microbial strain or genetic marker associated specifically with the gut microbiota of a particular host species.

Microbial source tracking encompasses a variety of phenotypic and genotypic methods used to identify a source-specific characteristic of a microorganism.[114] For example, researchers have explored the use of antibiotic resistance or carbon utilization profiles to match bacteria to their source.[115–117] Genetic fingerprinting techniques (including ribotyping and repetitive extragenic palindromic-PCR) have likewise been applied to match bacteria to sources of contamination.[118–120] These types of approaches are similar to the fingerprinting and DNA-typing techniques used in forensic science.

In microbial source tracking, approaches that rely on matching characteristics from an environmental microbe to one from a known source (*e.g.* cow feces) are termed 'library-dependent,' and a number of issues have been recognized with this approach.[121–123] Depending on the characteristic that is being used to match isolates, methods are sometimes too discriminatory, but at other times not discriminatory enough. For example, penicillin-resistant *E. coli* is almost ubiquitous in the environment from both human and animal wastes: identifying this antibiotic-resistance pattern in a water isolate does not help distinguish among its potentially numerous sources.[124]

On the other hand, many of the genetic fingerprinting methods may be too discriminatory. Hundreds or thousands of isolates from a single source may be necessary to represent strain variability, or else a water isolate may not find a match in the library, *i.e.* that database of known sources. Some 'cosmopolitan' bacterial strains have also been identified from multiple sources, so their identification in the environment does not help to distinguish sources.[125] There is also evidence that libraries are geographically and temporally specific. That means that a library built for one location may not be applicable to another. Furthermore, that library will likely need to be updated over time to account for genetic drift and other variations in population composition. When utilizing library-dependent methods, or any fecal source tracking technique, it is important for managers to carefully consider proper controls, method performance criteria and field-study design.[125]

Given the inherent limitations to using library-dependent source tracking approaches, library-independent methods have been developed that are much more simple and practical to use.[126] These methods rely on identification of a microbial species or genetic target that is generally known to be associated with a particular source. No database of microbial characteristics needs to be built or maintained. Many of the alternative indicators mentioned in the previous section also demonstrate potential as source tracking markers.

Cultivation of certain microbial species, including sorbitol-fermenting bifidobacteria[127–129] and *Rhodococcus coprophilus*,[130,131] has been applied to help track fecal contamination from humans and grazing animals, respectively. Bacteriophages have also been widely studied for their potential to discriminate sources of fecal pollution.[132–134] To date, the most promising bacteriophages proposed for microbial source tracking include coliphages and phages that infect *Bacteroides* (including phages of *B. fragilis, B. thetaiotaomicron, B. ruminicola* and *B. ovatus*). The F+ male-specific RNA coliphage group has proven particularly useful in distinguishing sources, although differential survival of the sub-types makes quantitation of sources problematic.[135,136]

More recently, identification of particular genes or DNA sequences has been proposed as more specific for sources of fecal contamination. One of the most popular source tracking targets in current use is the anaerobic bacterium, Bacteroidales.[137–139] There is evidence that certain strains have co-evolved with their hosts, so that the genetic signature of bacteria originating from non-human animals would be different from that originating from humans. Similarly, researchers have explored detection of the *esp* gene in the fecal indicator *Enterococcus*.[140,141] An additional and alternative anaerobic target proposed to track sewage pollution includes the *nif*H gene of *Methanobrevibacter smithii*.[142] *Methanobrevibacter smithii* is the most common methanogen in the human gut, occurring in concentrations up to 10^7–10^{10} per gram dry weight and its environmental persistence has been shown to exceed 24 days.[143]

No one approach or measure has emerged as an absolute for distinguishing sources of fecal pollution and no gold standard exists for this field of study.[133,134,144] Instead, a 'toolbox' of methods is being assembled.[50,67,145,146] Various methods could be used from this toolbox, depending on the needs of an individual situation. Another strategy is a 'weight of evidence' approach. For example, a large study conducted in the European Union, focused on comparing methods, concluded that information combined from multiple methods increases the likelihood of correct identifications.[134] This finding has since been confirmed in additional studies.[147]

5 Molecular Methods: A Revolution in Detection Technologies

Improved detection methods are needed in order to evaluate many of the alternative indicators currently proposed and many of the pathogens of concern in marine waters. The most commonly used molecular methods for analysis of water require the extraction, amplification and identification of nucleic acids including DNA and RNA.[148] These molecular techniques include the use of the polymerase chain reaction (PCR) and offer powerful tools for detecting and tracking microbial contaminants. Furthermore, these techniques tend to be more specific and rapid than traditional culture-based detection of microorganisms, features that can be critically important for decision making.[149]

Despite the promise of molecular detection methods, there are challenges that must be overcome in order to realize the potential of molecular methods for the marine environment. Application of these techniques for the detection of many enteric pathogens is severely limited by the lack of suitable concentration methods for initial isolation of microbes from large volumes of seawater or from the tissue of fish or shellfish.[149] Pathogens, especially enteric viruses, are usually found in dilute quantities. Therefore, large volumes (up to 1000 l) may need to be concentrated to a small volume of a few milliliters to allow for detection *via* culture or molecular methods. This process is critical, although logistically challenging, because many of the pathogens of interest have low infectious doses. Sometimes it only takes one virus or bacterium to cause illness to a human coming in contact directly or indirectly with the ocean. Innovative concentration techniques able to simultaneously concentrate different classes of microbes are helping to overcome this challenge.[150,151]

Increasing filtration volumes may be counterproductive, however, due to the increased capture of environmental inhibitors. It is problematic to isolate microbes from complex samples without also concentrating compounds that inhibit subsequent molecular analyses (*e.g.* humic acids). Recovery efficiencies tend to be low, both for concentration protocols as well as the nucleic acid extraction and purification steps that typically follow. The low recoveries decrease overall detection sensitivity and increase chances that organisms of public health concern could go undetected when they are present in low concentrations. Challenges associated with molecular detection can be divided into three broad steps: concentration, extraction and detection (see Figure 9). Inhibition and loss of recovery resulting from each step need to be quantified as part of any standardized protocol.[15,50,67]

Many of the challenges associated with molecular detection of microbes from environmental samples are being actively addressed in research laboratories, with notable successes. For example, in addition to typical positive controls, molecular detection assays are now routinely incorporating competitive and non-competitive internal controls, depending on whether they amplify with the same primers as the target nucleic acid.[152,153] Inhibition controls, most notably the use of salmon DNA to both quantify inhibition of a PCR assay and as an extraction control, are also becoming routine.[154] These internal controls are allowing for accurate quantitation of reaction inhibition, thus facilitating calibration of quantitative assays. Extraction controls are also being more aggressively pursued, with a better understanding that the most effective protocol or commercial extraction kit currently depends on the target of interest and the type of sample being processed. Challenges with sample concentration are in progress, with various researchers trying different combinations of commercially available filtration membranes and elution buffers. However, there are multiple reports of high efficiency filtration protocols.[155,156]

Detecting Microbes from Environmental Samples

Sample Concentration

Concentration Issues:
- Need to improve recovery efficiencies
- Choice of membrane - need to trap targets and prevent clogging
- Co-Concentration of PCR inhibitors
- Recovery of microbial targets (e.g. choosing an elution buffer and volume)
- Approaches other than filtration need to be better developed

Nucleic Acid Extraction

Extraction Issues:
- Co-Extraction of PCR inhibitors
- Controls needed to quantify loss of nucleic acids
- Need better reproducibility of extraction efficiencies
- Most effective kit or protocol will likely depend on target of interest and type of sample being processed

Nucleic Acid Detection

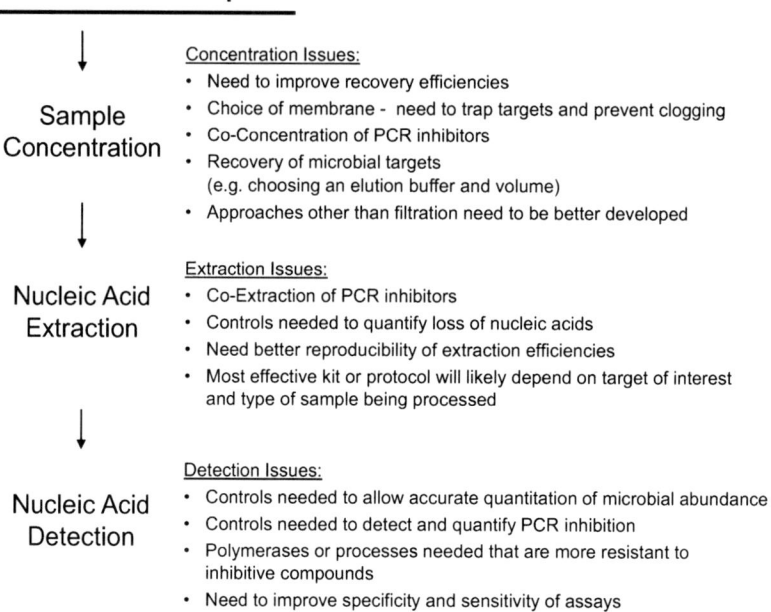

Detection Issues:
- Controls needed to allow accurate quantitation of microbial abundance
- Controls needed to detect and quantify PCR inhibition
- Polymerases or processes needed that are more resistant to inhibitive compounds
- Need to improve specificity and sensitivity of assays

Figure 9 Technological challenges associated with molecular detection of microorganisms from marine samples.

6 Epidemiological Studies: Linking Microbial Measures to Human Health

Epidemiology studies are used to link measures of microbial water quality to health effects associated with recreational water exposure (see Table 2) and serve as the basis for water quality criteria thresholds. Pruss[68] and Wade[69] reviewed epidemiological studies evaluating health effects from exposure to recreational water. In general, most studies reported a dose-related increase in swimming-related illnesses associated with increases in fecal indicator bacteria. The indicator organisms that reportedly associated best with the health outcomes were enterococci/fecal streptococci for marine and freshwater, and *E. coli* for freshwater. Of note, the majority of these studies were conducted at temperate beaches with known human sources of fecal contamination. That is, the majority of studies have focused on beaches which have a known point source (such as a sewage outfall) as more illnesses are likely to result from such a situation and the exposure is easier to ascertain.

Increasingly, particularly in developed nations regulating point sources of pollution, beaches may also be impacted by non-point sources.[15,50,67] The US EPA currently estimates that 60–80% of the impaired waters are polluted due to non-point source inputs. These are sources that may include storm water

Table 2 Summary of epidemiological studies linking microbial measures of water quality to human health outcomes. (Adapted and updated from Elmir).[178]

Author(s)	Study Location	Study Design & Indicators	Findings	Conclusions
Cabelli et al. (1972–1979),[179] sponsored by the USEPA.	Conducted at three beaches in the USA: New York City, NY; Lake Pontchartrain, New Orleans, LA; and Boston Harbor, MA.	Prospective cohort studies, approximately 26 686 total usable responses from all beaches over the 3-year studies. Enterococci, E. coli, Klebsiella, Enterobacter, Citrobacter, Total coliforms, C. Perfringens, P. aeruginosa, fecal coliforms, A. hydrophila, V. parahaemolyticus and Staphylococci were the indicators used for those studies.	Fecal coliforms, the indicator originally recommended in 1986 by the Federal Water Pollution Control Administration of the Department of Interior, showed less correlation to swimming-associated gastroenteritis than some other indicator organisms. E. coli showed strong correlation in fresh waters, whereas Enterococci showed strong correlation in both in fresh and marine waters.	The strong correlation may be a result of the survivability of the indicator organisms in the environment being similar to the survivability of the pathogens of concern. And, enterococci's resistance to environmental factors, particularly saline environments, enhancing its ability as a suitable indicator for marine waters.
Fattal et al. (1987)[180]	At three beaches marine waters with different water qualities, Tel-Aviv Israel.	Prospective cohort E. coli, fecal coliforms and enterococci were used to evaluate the microbial water quality.	Out of the three indicator-microbes tested, enterococci were the best indicator to predict GI illnesses among swimmers.	This finding agreed with the EPA epidemiological studies conducted by Cabelli et al.[179] in marine waters.
Cheung et al. (1990)[181]	At nine of the polluted (human waste discharge) beaches (marine waters), Hong Kong.	Prospective cohort 19 000 individuals participated in the study. Nine microbial indicators were used to evaluate the water quality; fecal coliforms, E. coli, Klebsiella spp., fecal streptococci, enterococci, staphylococci, Pseudomonas aeruginosa, Candida albicans and total fungi.	The strongest correlation between swimming-related health effects and an indicator density was between E. coli and highly credible gastrointestinal (HCGI) symptoms.	This finding does not agree with the EPA epidemiological studies conducted by Cabelli et al.[179] in marine waters.

Table 2 Continued.

Author(s)	Study Location	Study Design & Indicators	Findings	Conclusions
Balarajan et al. (1991)[182]	United Kingdom	Prospective cohort study that described the health risks related with exposure (wading, swimming, surfing and diving) to marine waters. 1883 individuals participated in the study. Information was not provided as to the parameters/indicator microbes used to evaluate the water quality at the study site.	The rate of enteric disease symptoms was significantly greater among bathers than non-bathers. The health risk for surfers/divers was approximately 1.4 times greater for swimmers and 1.5 times for waders.	The increase or decrease in health risk was concluded to be a function of type and degree of exposure.
von Schirnding et al. (1992)[183]	Prospective cohort at two beaches off the Atlantic coast of South Africa.	733 individuals participated in the study. One of the beaches was relatively clean; the other was considered to be moderately polluted due to failing septic tank systems and water runoff. Enterococci, fecal coliforms, coliphages and staphylococci were among the indicator microbes tested.	It was reported that there was a considerable increase in GI illness rates among swimmers than non-swimmers at the moderately polluted beach as oppose to the relatively clean beach.	It was concluded that there is increase in health risks among individuals exposed to polluted waters in comparison with individuals exposed to moderately polluted waters.
Corbett et al. (1993)[184]	Prospective cohort at the beaches (marine waters) in Sydney, Australia.	Conducted a study to assess the swimming-related illnesses, 2869 individuals participated in the study. Only fecal coliforms and fecal streptococci were used to measure the microbial quality of the waters.	It was found that for individuals who swam for more than 30 minutes, their risk of reporting GI symptoms increased by 4.6 times more than those who swam less than 30 minutes	This study showed similar results with the EPA beach water studies in that increasing GI illness rates were not associated with increasing fecal coliforms densities.

Kay et al. (1994)[78]	Randomized trial Beaches in the United Kingdom.	1112 individuals participated in the study, of which 512 were assigned to the swimmers group. The study was a randomized controlled epidemiological study. The microbial water quality was tested using total and fecal coliforms, fecal streptococci, total staphylococci and *Pseudomonas aeruginosa*.	Results of the study indicated that GI illness rates among swimmers were appreciably greater than non-swimmers. Out of the 4 indicator microbes, fecal streptococci were the best predictor for GI illness symptoms.	This finding agreed with the EPA epidemiological studies conducted by Cabelli et al.[179] in marine waters.
Fujioka et al. (1994)[185]	Prospective cohort, Hawaii, USA.	Individuals participating were classified in three distinct groups: non-swimmers, swimmers who did not swallow water and swimmers that did swallow water.	Study did not find any associations between the five microbial indicators (fecal coliforms, *E. coli*, enterococci, bacillus spores and *Clostridium perfringens*) analyzed and human health effects.	This finding does not agree with the EPA epidemiological studies conducted by Cabelli et al.[179] in marine waters.
Kueh et al. (1995)[186]	Prospective cohort in Hong Kong.	Analyzed the bacteriological aspect of water quality and examined how physicochemical parameters such as air, water temperature and turbidity may contribute to changes in microbial count and therefore bathing-related illness. Samples were analyzed for three bacterial indicators (*E. coli*, fecal coliforms and staphylococci) and seven pathogenic bacteria (*Aeromonas* spp., *Clostridium perfringens*, *Vibrio cholerae*, *V. parahaemolyticus*, *V. vulnificus*, *Salmonella* spp. and *Shigella* spp.).	In this study *Clostridium perfringens* and *Aeromonas* spp. showed a significant correlation with GI and HCGI symptoms, while *V. cholerae* and *V. parahaemolyticus* were best associated with GI but not HCGI symptoms.	Swimmers were, in general, two to three times more likely to develop illnesses than non-swimmers (swimmers only included those who wet their faces). The study also showed a strong correlation between water turbidity and GI and HCGI symptoms.

Table 2 Continued.

Author(s)	Study Location	Study Design & Indicators	Findings	Conclusions
Pruss (1998)[68]	Literature Review. The majority of these studies were conducted in the USA and UK, with few studies evaluated in tropical marine recreational waters.	Reviewed all significant existing epidemiological studies on the health effects from exposure to recreational water.	The indicator organisms that correlated best with the health outcomes were enterococci/fecal streptococci for marine and freshwater and *E. coli* for freshwater.	Review found that most studies reported a dose-related increase of health risk in swimmers with an increase in the indicator bacteria count in recreational water.
Fleisher *et al.* (1996)[79]	Randomized trial on 4 separate United Kingdom beaches during the summers of 1989 to 1992.	This particular study focused on how domestic sewage contamination pollutes marine waters and affects public health.	The results showed that the rates of illness (gastroenteritis, acute febrile respiratory illness, and eye and ear infections) among bathers were statistically significantly higher than for non-bathers.	The study showed a dose-response relationship between exposure and contaminated waters (among the bathers cohort, 34.4% to 65.8% of the adverse health conditions reported were considered a direct result of bathing in sewage-contaminated marine waters).
Haile *et al.* (1999)[57]	Prospective cohort, Santa Monica Bay, Los Angeles, USA.	Over 22 000 persons were interviewed 9 days after their facial immersion exposure to recreational beach waters concerning their symptoms. From the 22 085 subjects interviewed, 17 253 fulfilled the eligibility criteria, 15 492 agreed to participate and from those 13 278 were contacted during follow-up.	An increased risk of adverse health outcomes associated with swimming in ocean water contaminated by untreated urban runoff was found with a significant dose-response relationship.	An increased risk of adverse health outcomes associated with swimming in ocean water contaminated by untreated urban runoff was found, with a significant dose-response relationship.

Prieto et al. (2001)[187]	Prospective cohort, north of Spain.	Established a cohort of 2774 persons on 4 beaches in the north of Spain with follow up of 1858 persons after 7 days from exposure for symptoms. Among those followed up, 135 (7.5%) experienced symptoms; visitors experienced symptoms more than residents and symptoms were higher among bathers, although not significantly	Gastrointestinal and skin symptoms correlated with total coliforms; an increased risk was observed with exposure to 2500 to 9999 total coliforms per 100 ml. This coliforms count was below the European Union mandatory limit, although above a newly proposed standard.	This finding does not agree with the EPA epidemiological studies conducted by Cabelli et al.[179] in marine waters.
Fleming et al. (2004)[188]	Prospective cohort, in South Florida, USA (Hobie and Crandon Beaches).	Conducted a prospective cohort epidemiological pilot study at 2 beaches, using multiple bacteria indicators (enterococci, total and fecal coliforms, E. coli and C. perfringens). Final study population consisted of 63 families with 208 individuals. An epidemiological questionnaire was used to evaluate illness vs. exposure.	No significant association between the number and the type of reported symptoms and the different sampling months or beach sites. There was a negative correlation between the number of bacteria indicators and the frequency reported by beachgoers. Results of the daily monitoring indicated that different indicators provided conflicting results concerning beach water quality.	Larger epidemiological studies with individual exposure monitoring are recommended to further evaluate these potentially important associations in sub-tropical recreational waters.
Nova Southeastern University (2001–2003)[189]	Prospective cohort, in South Florida, USA (Hobie, Hollywood and Fort Lauderdale).	A voluntary beach questionnaire on three beaches was administered. Out of 10 000 surveys handed out only 892 experimental forms and 609 control forms were returned. Symptoms to be reported GI, upper respiratory, dermatological and constitutional.	Beach questionnaire didn't show clear signs of symptoms in the recreational population in comparison with the control population. Questionnaire return was low: around 10%.	A future and more comprehensive epidemiological study may be warranted.

Table 2 Continued.

Author(s)	Study Location	Study Design & Indicators	Findings	Conclusions
Colford et al. (2007)[71]	Prospective cohort in California, USA (6 popular public beaches in Mission Bay).	A cohort epidemiology study to evaluate the relationships between traditional indicators (enterococci, fecal coliforms and total coliforms) and swimming-related illnesses. Nearly 8800 participants were recruited for the study.	Only skin rash and diarrhea were consistently significantly elevated in swimmers compared to non-swimmers, especially among children 5 to 10 years old. The risk of illness was uncorrelated with levels of traditional water quality indicators. And the State water quality standards were not predictive of swimming-related illnesses.	Traditional fecal indicators were ineffective predictors of health effects and there is a need for further evaluation of traditional indicators at beaches where non-point sources are the dominant fecal contributors.
Wade et al. (2008)[190]	Prospective cohort at 4 freshwater Great Lakes beaches (USA) with point source pollution.	Prospective cohort with 21 015 interviews and 1359 water samples tested for *Enterococcus* using qPCR and membrane filtration.	*Enterococcus* QPCR cell equivalents (CEs) were positively associated with GI, and more strongly so among children aged <11. Weaker association between GI and *Enterococcus* by membrane filtration.	Measurement of the indicator bacteria *Enterococci* in recreational water using a rapid QPCR method predicted swimming-associated GI illness at freshwater beaches polluted by sewage

Fleisher et al. (2010)[72]	Randomized trial of a sub-tropical non-point source beach in South Florida, USA.	Randomized trial of a sub-tropical non-point source beach with 1300 participants. Samples analyzed for enterococci by membrane filtration, chromogenic substrate, qPCR, *S. aureus* and MRSA by membrane filtration. Bacteroides dog and human (UCD, HF8) by qPCR, Catellicoccus gull by qPCR.	Bathers were more likely to have GI, acute febrile respiratory illness and skin illness than non-bathers. Evidence of a dose–response relationship between skin illnesses and enterococci exposure.	Bathers may be at increased risk of several illnesses relative to non-bathers, even in the absence of any known source of domestic sewage impacting the recreational marine waters. No dose–response relationship between gastroenteritis and increasing exposure to enterococci, even though many current water-monitoring standards use gastroenteritis as the major outcome illness.

runoff, animal inputs, sand re-suspension and possibly human bather shedding. A safe assumption to make is that almost all, if not all, beaches are impacted by one or more non-point source of indicator organisms. Whether these indicator organisms track pathogens or are just due to regrowth of the particular indicator organism remains an open question.

To date, only a few large-scale epidemiological studies have been conducted at non-point source beaches. One of these studies was conducted in Mission Bay, CA, USA.[71] In this study, there were statistically significant increases in the risk for illness detected between bathers and non-bathers: odds ratio (OR) = 1.4 (95% confidence interval: 1.0–1.8) for skin rashes and 2.3 (1.6–3.2) for diarrhea for any water contact. Increases were also noted for those that swallowed water: 2.1 (1.4–2.0) for skin rash, 1.9 (1.3–2.7) for diarrhea and 1.7 (1.2–2.3) for eye irritation. An association was detected between illness (GI illness, nausea, cough and fever) and F+ (male-specific) coliphage. However, the authors interpreted this finding with caution since a low number of participants were exposed to water when F+ coliphage was detected.[71] Enterococci (as well as other indicator microbes) did not show a significant correlation with health risk in this study.

The second large-scale epidemiological study at a non-point source beach was conducted by Haile et al.[157] who investigated a beach that was impacted by runoff from storm drains. The investigators of this study detected a statistically significant greater risk of upper respiratory and GI illness as bathers swam closer to the runoff storm drains, and therefore concluded that there may be increased risk of swimming in runoff-contaminated waters. In this instance, the research did detect an association between illness and total coliforms, as well as with enteric viruses.

The two studies mentioned above in non-point source beaches were prospective cohort studies in which investigators interviewed individuals who decided to go into the water and compare them to individuals who decided to stay on the beach. This differs from the design of the third epidemiological study, the 'BEACHES' Study, conducted in Miami, Florida,[72,80] based on prior studies performed in the UK by Fleisher and Kay (see Table 2).[74] This study was the first randomized control exposure epidemiological study to be conducted in the United States, and the first to be conducted anywhere in an established non-point source environment. The study design involved randomly assigning healthy adults who report regularly bathing (*i.e.* wading or swimming) in South Florida to either a 'bather' or 'non-bather' group. The primary advantages of randomly assigning individuals who have a history of bathing (rather than comparing historic bathers to non-bathers) to either a bather or non-bather group are to minimize the bias and misclassification contained in all the prior cohort designs involving the health effects of bathing in contaminated recreational waters. Again this causes a possible bias in all prospective cohort studies of recreational water associated illness published to date.[74] The results of the BEACHES study demonstrated an increase in reported GI illnesses, skin ailments and acute febrile respiratory illness in bathers when compared to non-bathers.[72,80] A dose–response relationship was observed for skin ailments, where skin ailments were significantly and positively associated with enterococci enumeration by membrane filtration (OR = 1.46 [0.97–2.21] per increasing log10 unit of enterococci exposure); however, there was no association between report of GI and respiratory diseases with enterococcus, even though report of these illnesses was increased. Of note, this study had each individual bather collect their own water sample for enterococcus, reflecting their own personal 'marine water milieu'; this method of environmental monitoring has not been replicated in any other study to date.

No consistent association has been established between health risk and specific pathogens in marine beach epidemiological studies. This may be because only a few studies have included both the analysis of pathogens and measurements of human health. Some studies, however, have looked at correlations between the pathogen *Staphylococcus aureus* and bather health. However, *S. aureus* seems to be correlated with bather density and not a specific pollution source[68] and hence would be a concern mainly when beaches are overcrowded. A notable exception to this simplification would be for autoinfection which would be a risk regardless of bather density.[32,80] In another study, a correlation between enteric virus concentrations in runoff storm drains and bather health in the vicinity of the drains was noted.[157] Several recently completed or ongoing epidemiology studies around the USA have included several pathogens, but a relationship with health and specific pathogen exposure in these studies has not yet been clearly defined. Although pathogen presence itself in marine waters where humans are in contact is a reason for concern, the lack of relationship with health hinders the full reliance on pathogen monitoring.[67] As discussed below, modeling of pathogens and fecal indicators in marine waters is also being proposed as an alternative to traditional, single grab-sample monitoring to evaluate health risks.

7 Modeling Pathogens in Marine Waters

Modeling complex systems is an iterative process that requires collaboration of diverse disciplines. There are many purposes for applying mathematical models to the study of aquatic pathogens, including: predicting disease outbreaks; testing control measures; determining the relative importance of factors (*i.e.* parameters); identifying gaps in data; generating hypotheses; and increasing the overall understanding of the system of interest. In addition, models can be used to conduct virtual experiments that are otherwise unethical or logistically impossible. For the marine environment, models have been employed to estimate public health risk for both indigenous (*i.e. Vibrio* spp.) and introduced (*i.e.* fecal indicator bacteria) microbial pathogens.

7.1 Modeling Aquatic Pathogens: The Example of Vibrios

There are at least 50 described species in the genus *Vibrio*,[158] almost half of which are pathogenic to plants or animals, including 12 species implicated in human diseases (Table 1 in Morris; ref. 159). The predominant human pathogens are *V. cholerae* (the etiological agent of cholera, *V. parahaemolyticus*, the leading cause of vibrio-associated GI illness in the USA) and *V. vulnificus* (which may cause GI illnesses and wound infections.[160] Modeling efforts have focused on *V. cholerae* which has been responsible for seven pandemics since 1817, including the most recent from 1961 to the present time.[36]

The development of mathematical models for cholera began with the publication of a simple model based on the 1973 cholera outbreak in the Mediterranean.[161] This model focused on two key variables: (1) the number of susceptible people, and (2) the concentration of *V. cholerae*. At the time, it was believed that *V. cholerae* could not reproduce in the environment, and consequently the model did not allow for bacterial growth (*i.e.* fecal contamination drove disease dynamics). Following the discovery of the relationship between *V. cholerae* and plankton[162] and the eventual paradigm shift,[36] a new model for cholera was developed by Codeço.[163] Codeço improved the earlier model by adding another variable (*i.e.* the number of infected people) and focusing on the role of the aquatic reservoir in driving endemic and epidemic cholera. In turn, Codeço's model inspired several other models, each taking a different approach. For example, Pascual *et al.*[164] added a fourth variable to account for variations in water volume. This addition allowed climate-scale, environmental variables (*e.g.* precipitation, flooding, drought, land use) to affect pathogen concentrations in addition to, and independent of, bacterial growth and decay, and illustrated the importance of factors that concentrate *V. cholerae* in driving cholera outbreaks.

Similarly, Koelle *et al.*[165] published a model and analysis demonstrating the importance of both intrinsic (*e.g.* waning and cross immunity in individuals) and extrinsic (*e.g.* precipitation, flooding and drought) drivers of cholera epidemics. Meanwhile, Hartley *et al.*,[166] following the discovery of a short-lived, but hyper-infectious, state of *V. cholerae*,[167] generated a model that mathematically included two 'types' of the bacteria (*i.e.* two state variables for high and low virulence), and thereby demonstrated the impact of a hyper-infectious state on the severity of an epidemic. Likewise, King *et al.*[168] produced a model that incorporated asymptomatic carriers (*i.e.* two state variables for mild and severe infections), defined as people with mild cases of cholera that do not seek medical attention but contribute to the environmental pathogen load. Their modeling results suggested that the role of the aquatic reservoir was in the endemicity of *V. cholerae*, and that asymptomatic carriers with rapidly waning immunity drive epidemic cholera outbreaks. Combining parts of the Hartley *et al.*[166] and King *et al.*[168] models, Miller Neilan *et al.*[169] incorporated both the hyper-infectious states of *V. cholerae* and the asymptomatic carriers to test control measures (*i.e.* rehydration/antibiotics, sanitation/hygiene and vaccination) and to optimize application of these intervention strategies. Their model (*i.e.* now with six variables for the number of susceptible, two classes of infected, number of recovered, and two classes of bacteria) demonstrated that the optimum balance of control measures depends on characteristics of the specific population, including the ratio of asymptomatic to symptomatic infections, the average recovery rate from cholera and the average length of immunity.

The next generation of models for cholera (and other microbial pollutant pathogens) will need to build upon these existing models, in order to continue to evaluate the relative contributions of transmission from an aquatic reservoir compared to human-to-human contact. In addition, other characteristics of the biology and ecology of *V. cholerae* (such as genetic strain variation, serotype conversions, viable but not culturable (VBNC) cells, environmental horizontal gene exchange and community ecology) may be incorporated to formulate more specific models. These models may then be useful in predicting the effects of global climate change on the spread of this and other waterborne diseases.[170]

7.2 Coupling Modeling and Remote Sensing

Validating mathematical models requires data which may come from laboratory experiments, field samples, hospitals, public health agencies and remote sensing *via* satellites. For aquatic environments, remotely sensed data have the potential to provide larger scale and longer time series information compared to *in situ* environmental testing. The satellite sensors do not detect microbial pathogens directly, but rather monitor environmental conditions (or proxies) that are conducive to the growth and persistence of specific pathogens (or indicators) in order to predict the relative risk to humans.[171] Examples of remotely sensed data relevant for waterborne diseases include: sea surface temperature, sea surface height, sea surface salinity, turbidity, chlorophyll, ocean color, distribution of wetlands and degree of flooding. Although remote sensing has been successfully used for tracking vector-borne diseases in terrestrial environments,[172] there are far fewer examples from aquatic environments, with the notable exceptions of *V. cholerae* and *V. parahaemolyticus*.

Lobitz *et al.*[173] used remotely sensed sea surface temperature and sea surface height (*i.e.* proxy for water volume) to demonstrate that the timing and spread of cholera outbreaks could be monitored and potentially used as an early warning system in Bangladesh. Sea surface temperatures affect the timing, duration, and concentrations of the local phytoplankton blooms which in turn impact the zooplankton blooms and, because of the association of *V. cholerae* and zooplankton, cholera outbreaks. Sea surface heights affect the inland incursion of plankton-laden saltwater and consequently, the exposure of humans to environmental *V. cholerae*. The authors provided examples of the importance of combining both metrics by highlighting data from 1993 when temperatures were within normal ranges, but water volume was exceptionally low and no cholera outbreak was observed until the sea surface heights increased later that year.

Another example of using ecological factors to monitor and predict the dynamics of an aquatic pathogen includes research on *V. parahaemolyticus*. Phillips *et al.*[174] investigated the usefulness of remotely sensed data for the prediction of *V. parahaemolyticus* concentrations inside oysters. They first demonstrated that using remotely sensed temperatures was equivalent to using *in situ* temperature measurements from fixed points, and then argued that the remotely sensed data had better resolution over a larger geographical area. Consequently, for the purposes of forecasting *V. parahaemolyticus* concentrations, the remotely sensed sea surface temperatures, coupled with remotely sensed salinity and chlorophyll measurements, were expected to provide the best data for their predictive model. Although the authors point out some problems with interruption in data due to cloud cover and ground clutter, overall the imagery provided a user-friendly format for the interpretation and management of risk associated with this type of exposure to *V. parahaemolyticus*.

7.3 Use of Models in Management: Fecal Indicator Bacteria

As discussed throughout this chapter, recreational bodies of water are typically monitored for FIB in order to estimate potential risk of GI illnesses in humans exposed to pathogens associated with sewage contamination. Consequently, water quality models tend to track FIB concentrations instead of any specific target (*e.g.* enteric viruses, *Shigella*, *Salmonella* or *Cryptosporidium*) and are designed to forecast FIB concentrations based on environmental parameters (such as precipitation, temperature, salinity, sunlight, turbidity, nutrients and hydrodynamics). For example, Boehm *et al.*[175] developed an analytical model to assess the relative contributions of dilution, grazing and inactivation of enterococci (a marine FIB) in determining the direction and length of shoreline adversely impacted by a point source of microbial pollution. Furthermore, Hou *et al.*,[176] using a multivariate regression model, demonstrated that predictive models were an improvement over the more commonly employed approach whereby managers rely on the previous days' bacterial concentrations, under the assumption that concentrations persist for at least one day. Forecasting models for water quality, similar to daily and weekly weather reports, would substantially support management decisions for recreational waters.

8 The Future of Beach Regulation

As knowledge increases regarding best management practices for beach regulation, new approaches and techniques are being explored to protect bathers from pathogens in marine waters. Although performing fairly well at beaches with point source pollution, the use of one indicator microbe to monitor and regulate water quality at marine beaches has many limitations (as outlined earlier). This is especially problematic for developed nations, where beaches are dominated by non-point source pollution. More flexible and comprehensive monitoring and pollution prevention techniques are being explored in order to provide a relatively equal, yet practical, protection level to bathers at various marine beaches.

The principle of flexibility has been determined to be a critical point in beach monitoring and regulation.[50,66,67,78] Beaches across the world can dramatically vary, from the different physical, chemical and biological beach conditions and pollution sources, to the preferences of the beach users, to the regulators' capabilities and monitoring/remediation resources. A single or even multiple set of microbiological standards may not adequately address this diversity. Flexibility is being explored through the provision of more customized programs that are targeted to the needs of the individual beach types. This may be established through international, regional or national regulatory agencies such as the World Health Organization (WHO), European Union (EU), African Union and US EPA, through developing general guidelines and recommendations summarizing scientific advancements in the area of recreational marine water regulation. These guidelines would serve as a one-stop resource which teams (consisting of regulators and scientists, as well as other concerned local parties with expertise in the specific local conditions) can use to strategize a local policy that should work best for them. These general guidelines would need to be frequently updated to include the latest tried and true advancements and recommendations.

One concern of the critics in providing too much flexibility to the management of individual beaches is that health protection may not be equal across the regulated region.[50,67] This may be addressed through a diverse team of expert scientists, regulators and activists under the regulatory agency which can review proposed beach regulation plans of the individual states/counties/beaches. Given the specificity of the beach, as well as the knowledge of minimal protection that should be provided, this panel may either approve the proposed plan or recommend certain modifications to develop criteria that are acceptable. Therefore, the baseline that is used is not a single microbiological criterion, which may give a false sense of equal protection (false since recommended indicators perform differently at different beaches), but rather a single competent team of reviewers.

Beach regulation plans should not be limited to identifying which microbe(s) should be monitored. The idea of a 'tool box' has been previously recognized by the US EPA.[50] In addition to microbial monitoring, such a toolbox should include factors such as sampling time, location, frequency and method; predictive modeling (both 'forecast' and 'nowcast'); information on how to communicate beach conditions to beachgoers; sanitary survey methods; source tracking/tracer study methods; and pollution prevention methods *via* infrastructure, beach sand re-nourishment, community education, animal control, *etc.*

Such an approach, although more challenging to regulate on a national or regional level, may be the direction of future marine beach regulation. However, other approaches which have already applied these concepts have been proposed and in some cases implemented. Two such approaches are those of the WHO and the EU.[65,66,75] In brief, the WHO approach is a significant improvement over traditional monitoring guidelines in that differences between individual beaches are taken into account; hence the concept of flexibility mentioned earlier. This is done by applying a monitoring criterion that is based on the susceptibility of the beach to fecal pollution.[65] Beaches which do not have any known human fecal pollution sources are allowed to contain higher levels of indicator microbes than ones which do have a known point source of human fecal pollution. Such an approach is beneficial in that it limits an overly conservative approach to beaches where increases in indicator microbe concentrations are most likely not due to any human fecal source, and hence not likely to impose a human health risk. This approach also limits an 'under conservative' approach by not allowing beaches with known human fecal sources to contain high levels of indicator microbes, as this is likely due to the point source and hence would result in a human health risk. One of the limitations of this approach, however, is that it is currently limited to the one traditional indicator microbe (enterococci) monitoring criterion for marine beaches.

The second approach, which is more comprehensive, is that of the European Union.[66,75] The EU directive concerning management of bathing waters has been in use in the EU since 2008. This approach takes into account several critical aspects such as regulating the sampling program (*e.g.* frequency of sampling, location of sampling, time of sampling, *etc.*), including algae as indicators of human health risk and, most importantly, the requirement of beach water profiles for all beaches in the EU. The beach water profile is a critical addition to regulation since it requires beaches to be assessed for physical, geographical and hydrological conditions, as well as potential pollution sources. These profiles are to be updated every two to four years depending on their history (*i.e.* whether determined to be rated: excellent, good, sufficient or poor).[66] One of the limitations of this approach is that the monitoring criterion used is not based on the beach water profile, and therefore does not contain the aspect of flexibility found in the WHO approach.[65]

In the USA, current regulation guidelines include monitoring for one indicator microbe (enterococci) with one allowable level. However, as the US EPA moves rapidly to establish new recreational marine water criteria to be released in 2012, the aspects mentioned above (as well as the approaches of the EU and WHO) may also be taken into account.[50,67]

References

1. L. E. Fleming, K. Broad, A. Clement, E. Dewailly, S. Elmir, A. Knap, S. A. Pomponi, S. Smith, H. Solo Gabriele and P. Walsh, *Mar. Pollut. Bull.*, 2006, **53**, 545.
2. L. C. Backer, J. K. Kish, H. M. Solo Gabriele and L. E. Fleming, in *Water and Sanitation-Related Diseases and the Environment: Challenges, Interventions and Preventive Measures*, ed. J. Selendy, John Wiley & Sons, New York, in press.
3. J. C. Cato, *Economics of Hazard Analysis and Critical Control Point (HACCP) Programmes,* FAO Fisheries Technical Paper. No. 381, FAO, Rome, 1998.
4. A. A. Butt, K. E. Aldridge and C. V. Sanders, *Lancet Infect. Dis.*, 2004, **4**, 201.

5. J. Tibbetts, *Environ. Health Perspect.*, 2004, **112**, A282.
6. M. Lynch, J. Painter, R. Woodruff and C. Braden, *MMWR Surveillance Summaries*, 2006, **55**, 1.
7. M. Iwamoto, T. Ayers, B. E. Mahon and D. L. Swerdlow, *Clin. Microbiol. Rev.*, 2010, **23**, 399.
8. R. L. Scharff, Health-related costs from foodborne illness in the US, http://www.producesafetyproject.org (accessed March 3, 2010).
9. E. P. Ralston, K.-P. H. and A. Beet, *Environ. Health*, submitted.
10. H. Shuval, *J. Water Health*, 2003, **1**, 53.
11. R. H. Dwight, L. M. Fernandez, D. B. Baker, J. C. Semenza and B. H. Olson, *J. Environ. Manage.*, 2005, **76**, 95.
12. S. Given, L. H. Pendleton and A. B. Boehm, *Environ. Sci. Technol.*, 2006, **40**, 4851.
13. E. Halliday and R. J. Gast, *Environ. Sci. Technol.*, 2011, **45**, 370.
14. J. R. Thompson, L. A. Marcelina and M. F. Polz, in *Oceans and Health: Pathogens in the Marine Environment*, ed. S. Belkin and R. R. Colwell, Springer, New York, 2005.
15. J. R. Stewart, R. J. Gast, R. S. Fujioka, H. M. Solo-Gabriele, J. S. Meschke, L. A. Amaral-Zettler, E. Del Castillo, M. F. Polz, T. K. Collier, M. S. Strom, C. D. Sinigalliano, P. D. Moeller and A. F. Holland, *Environ. Health*, 2008, **7**(Suppl 2), S3.
16. J. A. Fuhrman, *Nature*, 1999, **399**, 541.
17. A. P. Wyn-Jones and J. Sellwood, *J. Appl. Microbiol.*, 2001, **91**, 945.
18. J. Baudart, K. Lemarchand, A. Brisabois and P. Lebaron, *Appl. Environ. Microbiol.*, 2000, **66**, 1544.
19. E. K. Lipp, S. A. Farrah and J. B. Rose, *Mar. Pollut. Bull.*, 2001, **42**, 286.
20. L. Cellini, A. Del Vecchio, M. Di Candia, E. Di Campli, M. Favaro and G. Donelli, *J. Appl. Microbiol.*, 2004, **97**, 285.
21. M. I. Van Dyke, V. K. Morton, N. L. McLellan and P. M. Huck, *J. Appl. Microbiol.*, 2010, **109**, 1053.
22. J. W. Santo Domingo and J. Hansel, in *Oceans and Human Health: Risks and Remedies from the Seas*, ed. P. J. Walsh, S. L. Smith, L. E. Fleming, H. M. Solo-Gabriele and W. H. Gerwick, Elsevier, Burlington, MA, 2008.
23. P. Anderson and T. Fenchel, *Limnol. Oceaogr.*, 1985, **30**, 198.
24. K. L. Jellison, H. F. Hemond and D. B. Schauer, *Appl. Environ. Microbiol.*, 2002, **68**, 569.
25. P. A. Conrad, M. A. Miller, C. Kreuder, E. R. James, J. Mazet, H. Dabritz, D. A. Jessup, F. Gulland and M. E. Grigg, *Int. J. Parasitol.*, 2005, **35**, 1155.
26. J. Plutzer, J. Ongerth and P. Karanis, *Int. J. Hygiene Environ. Health*, 2010, **213**, 321.
27. R. D. Adam, *Microbiol. Rev.*, 1991, **55**, 706.
28. R. Fayer and J. M. Trout, in *Oceans and Health: Pathogens in the Marine Environment*, ed. S. Belkin and R. R. Colwell, Springer, New York, 2005.
29. M. Solic and N. Krstulovic, *Mar. Pollut. Bull.*, 1994, **28**, 696.
30. N. Charoenca and R. S. Fujioka, *Water Sci. Technol.*, 1995, **31**, 11.

31. S. M. Elmir, M. E. Wright, A. Abdelzaher, H. M. Solo-Gabriele, L. E. Fleming, G. Miller, M. Rybolowik, M. T. Peter Shih, S. P. Pillai, J. A. Cooper and E. A. Quaye, *Water Res.*, 2007, **41**, 3.
32. G. Gabutti, A. DeDonno, F. Bagordo and M. T. Montagna, *Mar. Pollut. Bull.*, 2000, **40**, 697.
33. O. Tolba, A. Loughrey, C. E. Goldsmith, B. C. Millar, P. J. Rooney and J. E. Moore, *Int. J. Hygiene Environ. Health*, 2008, **211**, 398.
34. R. S. Fujioka and T. M. Unutoa, *Wat. Sci. Technol.*, 2006, **54**, 169.
35. A. Bosch, R. X. Abad and R. M. Pinto, in *Oceans and Health: Pathogens in the Marine Environment*, ed. S. Belkin and R. R. Colwell, Springer, New York, 2005.
36. R. R. Colwell, *Science*, 1996, **274**, 2025.
37. A. A. Elmanama, M. I. Fahd, S. Afifi, S. Abdallah and S. Bahr, *Environ. Res.*, 2005, **99**, 1.
38. C. A. Ristori, S. T. Iaria, D. S. Gelli and I. N. Rivera, *Int. J. Environ. Health Res.*, 2007, **17**, 259.
39. J. O. Falkinham, *J. Appl. Microbiol.*, 2008, **107**, 356.
40. D. Novakova, P. Svec and I. Sedlacek, *Lett. Appl. Microbiol.*, 2009, **48**, 289.
41. L. Vezzulli, E. Pezzati, M. Moreno, M. Fabiano, L. Pane and C. Pruzzo, *Microbial Ecol.*, 2009, **58**, 808.
42. CDC Factsheet for Francisella tularensis, http://www.bt.cdc.gov/agent/tularemia/faq.asp.
43. G. S. Visvesvara, H. Moura and F. L. Schuster, *FEMS Immunol. Med. Microbiol.*, 2007, **50**, 1.
44. B. M. Hsu, C. L. Lin and F. C. Shih, *Water Res.*, 2009, **43**, 2817.
45. F. L. Schuster and G. S. Visvesvara, *Vet. Parasitol.*, 2004, **126**, 91.
46. G. Greub and D. Raoult, *Clin. Microbiol. Rev.*, 2004, **17**, 413.
47. R. J. Gast, D. M. Moran, M. R. Dennett, W. A. Wurtsbaugh and L. A. Amaral-Zettler, *J Water Health*, in press.
48. R. Philipp, *Rev. Med. Microbiol.*, 1991, **2**, 208.
49. S. E. Kidd, Y. Chow, S. Mak, P. J. Bach, H. Chen, A. O. Hingston, J. W. Kronstad and K. H. Bartlett, *Appl. Environ. Microbiol.*, 2007, **73**, 1433.
50. USEPA, *Report of the Experts Scientific Workshop on Critical Research Needs for the Development of New or Revised Recreational Water Quality Criteria. EPA-823-R-07-006*, Office of Water, Washington, DC, 2007.
51. M. Schaechter and B. I. Eisenstein, in *Mechanisms of Microbial Disease*. ed. M. Schaechter, N. C. Engleberg, B. I. Eisenstein and G. Medoff, Lippincott, Williams and Wilkins, Philadelphia, PA, 3rd edn, 1999.
52. C. Pruzzo, A. Huq, R. R. Colwell and G. Donelli, in *Oceans and Health: Pathogens in the Marine Environment*, ed. S. Belkin and R. R. Colwell, Springer, New York, 2005.
53. A. DePaola, J. L. Nordstrom, J. C. Bowers, J. G. Wells and D. W. Cook, *Appl. Environ. Microbiol.*, 2003, **69**, 1521.
54. R. Y. Kong, S. K. Lee, T. W. Law, S. H. Law and R. S. Wu, *Water Res.*, 2002, **36**, 2802.

55. S. J. Ho Sui, A. Fedynak, W. W. Hsiao, M. G. Langille and F. S. Brinkman, *PLoS One*, 2009, **4**, e8094.
56. L. W. Sinton, in *Oceans and Health: Pathogens in the Marine Environment*, ed. S. Belkin and R. R. Colwell, Springer, New York, 2005.
57. A. C. Wright, R. T. Hill, J. A. Johnson, M. C. Roghman, R. R. Colwell and J. G. Morris, Jr., *Appl. Environ. Microbiol.*, 1996, **62**, 717.
58. M. M. Lyons, Y. Lau, W. E. Carden, J. E. Ward, S. B. Roberts, R. Smolowitz, J. Vallino and B. Allam, *EcoHealth*, 2007, **4**, 406.
59. M. M. Lyons, J. E. Ward, H. Gaff, R. E. Hicks, J. M. Drake and F. C. Dobbs, *Aquatic Microbial Ecol.*, 2010, **60**, 1.
60. J. E. M. Filho, R. M. Lopes, I. N. G. Rivera and R. R. Colwell, *J. Plankton Res.*, 2010, **33**, 51.
61. J. K. Parker, D. McIntyre and R. T. Noble, *Water Res.*, 2010, **44**, 4186.
62. R. M. Maier, I. L. Pepper and C. P. Gerba, *Environmental Microbiology*, Academic Press, San Diego, CA, 2nd edn, 2009.
63. V. Cabelli, *Health Effects Criteria for Marine Recreational Waters*, EPA-600/1-80-031, US Environmental Protection Agency, Cincinnati, OH, 1983.
64. USEPA, *Method 1600: Membrane Filter Test Method for Enterococci in Water*, EPA-821-R-02-022, Washington, D.C., 2002.
65. WHO, *Guidelines for Safe Recreational Water Environments. Volume 1 Coastal and Fresh Waters*, Geneva, Switzerland, 2003.
66. EC, *Directive 2006/7/EC of the European Parliament and of the Council of 15 February 2006 Concerning the Management of Bathing Water Quality and Repealing Directive 76/160/EEC*, 2006, pp. 37–51.
67. A. B. Boehm, N. J. Ashbolt, J. M. Colford, Jr., L. E. Dunbar, L. E. Fleming, M. A. Gold, J. A. Hansel, P. R. Hunter, A. M. Ichida, C. D. McGee, J. A. Soller and S. B. Weisberg, *J. Water Health*, 2009, **7**, 9.
68. A. Pruss, *Int. J. Epidemiol.*, 1998, **27**, 1.
69. T. J. Wade, N. Pai, J. N. Eisenberg and J. M. Colford, Jr., *Environ. Health Perspect.*, 2003, **111**, 1102.
70. G. F. Craun, R. L. Calderon and T. J. Wade, *J. Water Health*, 2004, **4**(Suppl 2), 3.
71. J. M. Colford, Jr., T. J. Wade, K. C. Schiff, C. C. Wright, J. F. Griffith, S. K. Sandhu, S. Burns, M. Sobsey, G. Lovelace and S. B. Weisberg, *Epidemiology*, 2007, **18**, 27.
72. J. M. Fleisher, L. E. Fleming, H. M. Solo-Gabriele, J. K. Kish, C. D. Sinigalliano, L. Plano, S. M. Elmir, J. D. Wang, K. Withum, T. Shibata, M. L. Gidley, A. Abdelzaher, G. He, C. Ortega, X. Zhu, M. Wright, J. Hollenbeck and L. C. Backer, *Int. J. Epidemiol.*, 2010, **39**, 1291.
73. J. M. Fleisher, *Water Pollut. Control Fed. J.*, 1991, **63**, 259.
74. J. M. Fleisher and D. Kay, *Mar. Pollut. Bull.*, 2006, **52**, 264.
75. D. Kay, J. Bartram, A. Pruss, N. Ashbolt, M. D. Wyer, J. M. Fleisher, L. Fewtrell, A. Rogers and G. Rees, *Water Res.*, 2004, **38**, 1296.
76. NRDC, *Testing the Waters 2010: A Guide to Water Quality at Vacation Beaches*, Natural Resources Defense Council, New York, 2010.

77. T. Shibata, H. M. Solo-Gabriele, L. E. Fleming and S. Elmir, *Water Res.*, 2004, **38**, 3119.
78. D. Kay, J. M. Fleisher, R. L. Salmon, F. Jones, M. D. Wyer, A. F. Godfree, Z. Zelenauch-Jacquotte and R. Shore, *Lancet*, 1994, **344**, 905.
79. J. M. Fleisher, D. Kay, R. L. Salmon, F. Jones, M. D. Wyer and A. F. Godfree, *Am. J. Public Health*, 1996, **86**, 1228.
80. C. D. Sinigalliano, J. M. Fleisher, M. L. Gidley, H. M. Solo-Gabriele, T. Shibata, L. R. Plano, S. M. Elmir, D. Wanless, J. Bartkowiak, R. Boiteau, K. Withum, A. M. Abdelzaher, G. He, C. Ortega, X. Zhu, M. E. Wright, J. Kish, J. Hollenbeck, T. Scott, L. C. Backer and L. E. Fleming, *Water Res.*, 2010, **44**, 3763.
81. M. E. Wright, H. M. Solo-Gabriele, S. Elmir and L. E. Fleming, *Mar. Pollut. Bull.*, 2009, **58**, 1649.
82. M. E. Wright, H. M. Solo-Gabriele, A. M. Abdelzaher, S. Elmir and L. E. Fleming, *Water Sci. Technol.*, 2011, **63**, 542.
83. A. B. Boehm, *Environ. Sci. Technol.*, 2007, **41**, 8227.
84. M. E. Wright, H. M. Solo-Gabriele, A. M. Abdelzaher, S. Elmir and L. E. Fleming, *Environ. Sci. Technol.*, in press.
85. R. S. Fujioka, H. H. Hashimoto, E. B. Siwak and R. H. Young, *Appl. Environ. Microbiol.*, 1981, **41**, 690.
86. H. M. Solo-Gabriele, M. A. Wolfert, T. R. Desmarais and C. J. Palmer, *Appl. Environ. Microbiol.*, 2000, **66**, 230.
87. R. L. Whitman, M. B. Nevers, G. C. Korinek and M. N. Byappanahalli, *Appl. Environ. Microbiol.*, 2004, **70**, 4276.
88. L. J. Beversdorf, S. M. Bornstein-Forst and S. L. McLellan, *J. Appl. Microbiol.*, 2007, **102**, 1372.
89. S. Weisberg, A. Dufour, F. Galt, M. Gold, M. Noble, R. Noble, E. Reichard, P. Roberts, J. Rose and D. Rosenblatt, *Huntington Beach Closure Investigation. Technical Review USCSG-TR-01-2000,* Sea Grant, University of Southern California, San Diego, CA, 2000.
90. D. W. Griffin, K. A. Donaldson, J. H. Paul and J. B. Rose, *Clin. Microbiol. Rev.*, 2003, **16**, 129.
91. W. Q. Betancourt and J. B. Rose, *Vet. Parasitol.*, 2004, **126**, 219.
92. J. C. Chang, S. F. Ossoff, D. C. Lobe, M. H. Dorfman, C. M. Dumais, R. G. Qualls and J. D. Johnson, *Appl. Environ. Microbiol.*, 1985, **49**, 1361.
93. T. R. Deetz, E. M. Smith, S. M. Goyal, C. P. Gerba, J. J. Vollet, Y. L. Tsai, H. L. Du Pont and B. H. Keswick, *Water Res.*, 1984, **18**, 567.
94. S. C. Jiang and W. Chu, *J. Appl. Microbiol.*, 2004, **97**, 17.
95. R. T. Noble and J. A. Fuhrman, *Hydrobiologia*, 2001, **460**, 175.
96. R. S. Fujioka and M. N. Byappanahalli, *Final Report Tropical Indicator Workshop,* Prepared for EPA, Office of Water by the Water Resources Research Institute, University of Hawaii, Manoa, 2001.
97. L. V. Holdeman, I. J. Good and W. E. Moore, *Appl. Environ. Microbiol.*, 1976, **31**, 359.
98. C. A. Kreader, *Appl. Environ. Microbiol.*, 1995, **61**, 1171.

99. A. M. Abdelzaher, W. M. E., C. Ortega, A. R. Hasan, T. Shibata, H. M. Solo-Gabriele, J. Kish, K. Withum, G. He, S. M. Elmir, J. A. Bonilla, T. D. Bonilla, C. J. Palmer, T. M. Scott, J. Lukasik, V. J. Harwood, S. McQuaig, C. D. Sinigalliano, M. L. Gidley, D. Wanless, L. R. W. Plano, A. C. Garza, X. Zhu, S. J. R., J. W. Dickerson, H. Yampara-Iquise, C. Carson, J. M. Fleisher and L. E. Fleming, *J. Water Health*, in press.
100. S. M. McQuaig, T. M. Scott, V. J. Harwood, S. R. Farrah and J. O. Lukasik, *Appl. Environ. Microbiol.*, 2006, **72**, 7567.
101. S. Bofill-Mas, S. Pina and R. Girones, *Appl. Environ. Microbiol.*, 2000, **66**, 238.
102. A. M. Abdelzaher, M. E. Wright, C. Ortega, H. M. Solo-Gabriele, G. Miller, S. Elmir, X. Newman, P. Shih, J. A. Bonilla, T. D. Bonilla, C. J. Palmer, T. Scott, J. Lukasik, V. J. Harwood, S. McQuaig, C. Sinigalliano, M. Gidley, L. R. Plano, X. Zhu, J. D. Wang and L. E. Fleming, *Appl. Environ. Microbiol.*, 2010, **76**, 724.
103. K. Rosario, E. M. Symonds, C. Sinigalliano, J. Stewart and M. Breitbart, *Appl. Environ. Microbiol.*, 2009, **75**, 7261.
104. R. T. Noble, S. M. Allen, A. D. Blackwood, W. Chu, S. C. Jiang, G. L. Lovelace, M. D. Sobsey, J. R. Stewart and D. A. Wait, *J. Water Health*, 2003, **1**, 195.
105. A. Hundesa, C. Maluquer de Motes, S. Bofill-Mas, N. Albinana-Gimenez and R. Girones, *Appl. Environ Microbiol.*, 2006, **72**, 7886.
106. N. Albinana-Gimenez, P. Clemente-Casares, S. Bofill-Mas, A. Hundesa, F. Ribas and R. Girones, *Environ. Sci. Technol.*, 2006, **40**, 7416.
107. K. A. Peeler, S. P. Opsahl and J. P. Chanton, *Environ. Sci. Technol.*, 2006, **40**, 7616.
108. C. Hagedorn and S. B. Weisberg, *Rev. Environ. Sci. Biotechnol.*, 2009, **8**, 275.
109. S. Seurinck, W. Verstraete and S. D. Siciliano, *Rev. Environ. Sci. Biotechnol.*, 2005, **4**, 19.
110. J. W. Santo Domingo, D. G. Bambic, T. A. Edge and S. Wuertz, *Water Res.*, 2007, **41**, 3539.
111. T. M. Scott, J. B. Rose, T. M. Jenkins, S. R. Farrah and J. Lukasik, *Appl. Environ. Microbiol.*, 2002, **68**, 5796.
112. E. E. Geldreich and B. A. Kenner, *J. Water Pollut. Control Fed.*, 1969, **41**(Suppl R336+).
113. S. Wuertz and J. Field, *Water Res.*, 2007, **41**, 3515.
114. C. L. Meays, K. Broersma, R. Nordin and A. Mazumder, *J. Environ. Manage.*, 2004, **73**, 71.
115. C. W. Kaspar, J. L. Burgess, I. T. Knight and R. R. Colwell, *Can. J. Microbiol.*, 1990, **36**, 891.
116. B. A. Wiggins, R. W. Andrews, R. A. Conway, C. L. Corr, E. J. Dobratz, D. P. Dougherty, J. R. Eppard, S. R. Knupp, M. C. Limjoco, J. M. Mettenburg, J. M. Rinehardt, J. Sonsino, R. L. Torrijos and M. E. Zimmerman, *Appl. Environ. Microbiol.*, 1999, **65**, 3483.
117. C. Hagedorn, J. B. Crozier, K. A. Mentz, A. M. Booth, A. K. Graves, N. J. Nelson and R. B. Reneau, Jr., *J. Appl. Microbiol.*, 2003, **94**, 792.

118. S. Parveen, K. M. Portier, K. Robinson, L. Edmiston and M. L. Tamplin, *Appl. Environ. Microbiol.*, 1999, **65**, 3142.
119. P. E. Dombek, L. K. Johnson, S. T. Zimmerley and M. J. Sadowsky, *Appl. Environ. Microbiol.*, 2000, **66**, 2572.
120. S. P. Myoda, C. A. Carson, J. J. Fuhrmann, B. K. Hahm, P. G. Hartel, H. Yampara-Lquise, L. Johnson, R. L. Kuntz, C. H. Nakatsu, M. J. Sadowsky and M. Samadpour, *J. Water Health*, 2003, **1**, 167.
121. P. G. Hartel, J. D. Summer, J. L. Hill, J. V. Collins, J. A. Entry and W. I. Segars, *J. Environ. Qual.*, 2002, **31**, 1273.
122. L. K. Johnson, M. B. Brown, E. A. Carruthers, J. A. Ferguson, P. E. Dombek and M. J. Sadowsky, *Appl. Environ. Microbiol.*, 2004, **70**, 4478.
123. V. J. Harwood, in *Microbial Source Tracking*, ed. J. W. Santo Domingo and M. J. Sadowsky, ASM Press, Washington, DC, 2007.
124. J. A. Gooch, J. R. Stewart, B. C. Thompson, L. F. Webster, S. M. Allen, L. M. Kracker, P. L. Pennington, M. H. Fulton and G. I. Scott, *Use of Three Microbial Source Tracking Methods to Analyze Shellfish Harvesting Waters in South Carolina*, Report Submitted to the South Carolina Department of Health and Environmental Control, Columbia, SC, 2004, p. 93.
125. D. M. Stoeckel and V. J. Harwood, *Appl. Environ. Microbiol.*, 2007, **73**, 2405.
126. K. G. Field, E. C. Chern, L. K. Dick, J. Fuhrman, J. Griffith, P. A. Holden, M. G. LaMontagne, J. Le, B. Olson and M. T. Simonich, *J. Water Health*, 2003, **1**, 181.
127. I. G. Resnick and M. A. Levin, *Appl. Environ. Microbiol.*, 1981, **42**, 433.
128. D. D. Mara and J. I. Oragui, *J. Appl. Bacteriol.*, 1983, **55**, 349.
129. S. C. Long, A. P. Catalina and J. D. Plummer, *Can. J. Microbiol.*, 2005, **51**, 413.
130. D. D. Mara and J. I. Oragui, *Appl. Environ. Microbiol.*, 1981, **42**, 1037.
131. M. G. Savill, S. R. Murray, P. Scholes, E. W. Maas, R. E. McCormick, E. B. Moore and B. J. Gilpin, *J. Microbiol. Methods*, 2001, **47**, 355.
132. J. Jofre, J. R. Stewart and W. Grabow, in *Microbial Source Tracking: Methods, Applications and Case Studies*, ed. C. Hagedorn, A. R. Blanch and V. J. Harwood, in press.
133. J. F. Griffith, S. B. Weisberg and C. D. McGee, *J. Water Health*, 2003, **1**, 141.
134. A. R. Blanch, L. Belanche-Munoz, X. Bonjoch, J. Ebdon, C. Gantzer, F. Lucena, J. Ottoson, C. Kourtis, A. Iversen, I. Kuhn, L. Moce, M. Muniesa, J. Schwartzbrod, S. Skraber, G. T. Papageorgiou, H. Taylor, J. Wallis and J. Jofre, *Appl. Environ. Microbiol.*, 2006, **72**, 5915.
135. G. M. Brion, J. S. Meschke and M. D. Sobsey, *Water Res.*, 2002, **36**, 2419.
136. J. Stewart-Pullaro, J. W. Daugomah, D. E. Chestnut, D. A. Graves, M. D. Sobsey and G. I. Scott, *J. Appl. Microbiol.*, 2006, **101**, 1015.
137. A. E. Bernhard and K. G. Field, *Appl. Environ. Microbiol.*, 2000, **66**, 4571.

138. L. K. Dick, A. E. Bernhard, T. J. Brodeur, J. W. Santo Domingo, J. M. Simpson, S. P. Walters and K. G. Field, *Appl. Environ. Microbiol.*, 2005, **71**, 3184.
139. O. C. Shanks, J. W. Domingo, J. Lu, C. A. Kelty and J. E. Graham, *Appl. Environ. Microbiol.*, 2007, **73**, 2416.
140. T. M. Scott, T. M. Jenkins, J. Lukasik and J. B. Rose, *Environ. Sci. Technol.*, 2005, **39**, 283.
141. R. L. Whitman, K. Przybyla-Kelly, D. A. Shively and M. N. Byappanahalli, *Environ. Sci. Technol.*, 2007, **41**, 6090.
142. C. Johnston, J. A. Ufnar, J. F. Griffith, J. A. Gooch and J. R. Stewart, *J. Appl. Microbiol.*, 2010, **109**, 1946.
143. J. A. Ufnar, S. Y. Wang, J. M. Christiansen, H. Yampara-Iquise, C. A. Carson and R. D. Ellender, *J. Appl. Microbiol.*, 2006, **101**, 44.
144. D. M. Stoeckel, M. V. Mathes, K. E. Hyer, C. Hagedorn, H. Kator, J. Lukasik, T. L. O'Brien, T. W. Fenger, M. Samadpour, K. M. Strickler and B. A. Wiggins, *Environ. Sci. Technol.*, 2004, **38**, 6109.
145. J. R. Stewart, R. D. Ellender, J. A. Gooch, S. Jiang, S. P. Myoda and S. B. Weisberg, *J. Water Health*, 2003, **1**, 225.
146. K. G. Field and M. Samadpour, *Water Res.*, 2007, **41**, 3517.
147. M. Gourmelon, M. P. Caprais, S. Mieszkin, R. Marti, N. Wery, E. Jarde, M. Derrien, A. Jadas-Hecart, P. Y. Communal, A. Jaffrezic and A. M. Pourcher, *Water Res.*, **44**, 4812.
148. J. Jofre and A. R. Blanch, *J. Appl. Microbiol.*, 2010, **109**, 1853.
149. R. T. Noble and S. B. Weisberg, *J. Water Health*, 2005, **3**, 381.
150. A. M. Abdelzaher, H. M. Solo-Gabriele, C. J. Palmer and T. M. Scott, *J. Environ. Qual.*, 2009, **38**, 2468.
151. V. R. Hill, A. L. Polaczyk, D. Hahn, J. Narayanan, T. L. Cromeans, J. M. Roberts and J. E. Amburgey, *Appl. Environ. Microbiol.*, 2005, **71**, 6878.
152. J. B. Gregory, R. W. Litaker and R. T. Noble, *Appl. Environ. Microbiol.*, 2006, **72**, 3960.
153. J. B. Gregory, L. F. Webster, J. F. Griffith and J. R. Stewart, *J. Virol. Methods*, 2011, **172**, 38.
154. R. A. Haugland, S. C. Siefring, L. J. Wymer, K. P. Brenner and A. P. Dufour, *Water Res.*, 2005, **39**, 559.
155. H. Katayama, A. Shimasaki and S. Ohgaki, *Appl. Environ. Microbiol.*, 2002, **68**, 1033.
156. H. A. Morales-Morales, G. Vidal, J. Olszewski, C. M. Rock, D. Dasgupta, K. H. Oshima and G. B. Smith, *Appl. Environ. Microbiol.*, 2003, **69**, 4098.
157. R. W. Haile, J. S. Witte, M. Gold, R. Cressey, C. McGee, R. C. Millikan, A. Glasser, N. Harawa, C. Ervin, P. Harmon, J. Harper, J. Dermand, J. Alamillo, K. Barrett, M. Nides and G. Wang, *Epidemiology*, 1999, **10**, 355.
158. J. G. Holt, N. R. Krieg, P. H. A. Sneath, J. T. Staley and S. T. Williams, *Bergey's Manual of Determinative Bacteriology*, Lippincot, Williams & Wilkins, Philadelphia, PA, 9th edn., 2000.

159. J. G. Morris, *Food Safety*, 2003, **37**, 272.
160. D. J. Grimes, C. N. Johnson, K. S. Dillon, A. R. Flowers, N. F. Noriea, 3rd and T. Berutti, *Microb. Ecol.*, 2009, **58**, 447.
161. V. Capasso and S. L. Paveri-Fontana, *Rev. Epidemiol. Sante Publique*, 1979, **27**, 121.
162. R. R. Colwell, J. Kaper and S. W. Joseph, *Science*, 1977, **198**, 394.
163. C. T. Codeco, *BMC Infect. Dis.*, 2001, **1**, 1.
164. M. Pascual, M. J. Bouma and A. P. Dobson, *Microbes Infect.*, 2002, **4**, 237.
165. K. Koelle, X. Rodo, M. Pascual, M. Yunus and G. Mostafa, *Nature*, 2005, **436**, 696.
166. D. M. Hartley, J. G. Morris, Jr. and D. L. Smith, *PLoS Med*, 2006, **3**, e7.
167. D. S. Merrell, S. M. Butler, F. Qadri, N. A. Dolganov, A. Alam, M. B. Cohen, S. B. Calderwood, G. K. Schoolnik and A. Camilli, *Nature*, 2002, **417**, 642.
168. A. A. King, E. L. Ionides, M. Pascual and M. J. Bouma, *Nature*, 2008, **454**, 877.
169. R. L. Miller Neilan, E. Schaefer, H. Gaff, K. R. Fister and S. Lenhart, *Bull. Math. Biol.*, 2010, **72**, 2004.
170. K. Koelle, *Clin. Microbiol. Infect.*, 2009, **15**(Suppl 1), 29.
171. J. Travis, *Science News*, 1997, **152**, 72.
172. L. R. Beck, B. M. Lobitz and B. L. Wood, *Emerg. Infect. Dis.*, 2000, **6**, 217.
173. B. Lobitz, L. Beck, A. Huq, B. Wood, G. Fuchs, A. S. Faruque and R. Colwell, *Proc. Natl. Acad. Sci. U. S. A.*, 2000, **97**, 1438.
174. A. M. Phillips, A. Depaola, J. Bowers, S. Ladner and D. J. Grimes, *J. Food Protect.*, 2007, **70**, 879.
175. A. B. Boehm, D. P. Keymer and G. G. Shellenbarger, *Water Res.*, 2005, **39**, 3565.
176. D. Hou, S. J. Rabinovici and A. B. Boehm, *Environ. Sci. Technol.*, 2006, **40**, 1737.
177. R. M. Maier, I. L. Pepper and C. P. Gerba, *Environmental Microbiology*, Academic Press, San Diego, CA, 1st edn., 2000.
178. S. Elmir, Development of a water quality model which incorporates non-point microbial sources, PhD Thesis, University of Miami, CA, 2006.
179. V. J. Cabelli, A. P. Dufour, L. J. McCabe and M. A. Levin, *Am. J. Epidemiol.*, 1982, **115**, 606.
180. B. Fattal, E. Peleg-Olevsky, T. Agursky and H. Shuval, *Chemosphere*, 1987, **115**, 606.
181. W. H. Cheung, K. C. Chang, R. P. Hung and J. W. Kleevens, *Epidemiol. Infect.*, 1990, **105**, 139.
182. R. Balarajan, V. Soni Raleigh, P. Yuen, D. Wheeler, D. Machin and R. Cartwright, *Br. Med. J.*, 1991, **303**, 1444.
183. Y. E. von Schirnding, R. Kfir, V. Cabelli, L. Franklin and G. Joubert, *S. Afr. Med. J.*, 1992, **81**, 543.

184. S. J. Corbett, G. L. Rubin, G. K. Curry and D. G. Kleinbaum, *Am. J. Public Health*, 1993, **83**, 1701.
185. R. Fujioka, K. Roll and D. Morens, *A Pilot Epidemiological Study of Health Risks Associated with Swimming at Kuhio Beach*, Hawaii Water Resources Research Center, 1994.
186. C. S. W. Kueh, T. Y. Tam, T. Lee, S. L. Wong, O. L. Lloyd, I. T. S. Yu, T. W. Wong, J. S. Tam and D. C. J. Bassett, *Water Sci. Technol.*, 1995, **31**, 1.
187. M. D. Prieto, B. Lopez, J. A. Juanes, J. A. Revilla, J. Llorca and M. Delgado-Rodriguez, *J. Epidemiol. Community Health*, 2001, **55**, 442.
188. L. E. Fleming, G. H. Solo, S. Elmir, T. Shibata, D. Squicciarini, W. Quirino, M. Arguello and G. Van de Bogart, *Florida J. Environ. Health*, 2004, **184**, 29.
189. T. D. Bonilla, K. Nowosielski, M. Cuvelier, A. Hartz, M. Green, N. Esiobu, D. S. McCorquodale, J. M. Fleisher and A. Rogerson, *Mar. Pollut. Bull.*, 2007, **54**, 1472.
190. T. J. Wade, R. L. Calderon, K. P. Brenner, E. Sams, M. Beach, R. Haugland, L. Wymer and A. P. Dufour, *Epidemiology*, 2008, **19**, 375.

Estuarine and Marine Pollutants

JAMES W. READMAN,* ENIKO KADAR, JOHN A. J. READMAN AND CARLOS GUITART

ABSTRACT

With the notable exception of the methyl mercury poisoning event in Minamata Bay, Japan, in the 1950s, chemical pollutants, unlike pathogens and toxic algal blooms, rarely cause hospitalisation or instant death. Effects with respect to human health are far more subtle and are typically chronic rather than acute. Consumption of contaminated seafood is the major route of uptake and has implications with respect to increasing aquaculture. Marine aerosols afford another route of exposure for man. Typically, the socio-economic factors are most importantly affected through loss of amenities, ecology and produce, leading to degradation of the environment and, for example, reductions in tourism. This chapter complements those dealing with microbial pollution and harmful algal blooms, and addresses priority pollutants, emerging contaminants presently under scrutiny (including nanoparticles) and plastics. It also discusses the problems associated with evaluating complex mixtures of contaminants to which biota (including humans) are usually exposed. Climate change implications and its effects on pollution are also investigated. Finally, future issues of concern are debated.

1 Context

At the outset we need to understand what we mean by marine pollution. For many years most marine scientists have adopted the definition developed by the

---

*Corresponding author

United Nations Group of Experts on the Scientific Aspects of Marine Pollution (GESAMP) which is: "the introduction by man, directly or indirectly, of substances or energy into the marine environment (including estuaries) resulting in such deleterious effects as harm to living resources, hazards to human health, hindrance to marine activities including fishing, impairment of quality for use of sea water, and reduction of amenities". Indeed, the UN Convention on the Law of the Sea largely follows the GESAMP definition: "the introduction by man, directly or indirectly, of substances or energy into the marine environment, including estuaries, which results or is likely to result in such deleterious effects as harm to living resources and marine life, hazards to human health, hindrance to marine activities, including fishing and other legitimate uses of the sea, impairment of quality for use of sea water and reduction of amenities". Clearly, these definitions of pollution are human orientated.

This chapter focuses on 'chemical' pollutants. Human health issues relating to this topic do not usually afford the same dramatic impacts that can occur with microbial infection and the rapid toxicological effects associated with poisons produced in toxic algal blooms (topics dealt with in other chapters of this book). These can produce a rapid decline in human health often resulting in ill health, hospitalisation and sometimes death. Effects from chemical pollutants generally tend to be more subtle with respect to human health, affording more chronic (longer term) illness with cumulative/delayed effects. The most direct pathway of exposure is through the consumption of seafood.

One notable exception to this general rule is that of Minamata Bay in Japan. During the 1950s, in this region the population was exposed to contaminants through the consumption of shellfish which resulted in substantial neurological disorders. It was initially unclear as to the cause of the illness, but in the 1960s it became apparent that it could be traced to mercury (and more specifically methylmercury) poisoning. Estimates vary as to how many people were affected, ranging from 700 to 4000. Of the 700 estimated to be affected, it has been reported that 40% died.[1] Subsequent research indicated that mercury was being released into the bay from industrial sources (including the manufacturing of vinyl chloride as a catalyst and the industrial production of acetaldehyde). Whilst methylation of the mercury in the sediments of the bay was found to contribute to the formation of the responsible toxin, it is also likely that a proportion of the bioaccumulative methylmercury was also emitted in the industrial effluents.

As noted, however, chemical contaminant effects are generally more subtle, affording the potential for longer-term bioaccumulative effects relating to genotoxins and endocrine disruption. Priority pollutants and emerging contaminants are also addressed within this text. As human numbers inevitably increase and with the proportion of populations favouring locations adjacent to the sea, the relative contribution of their wastes finding their way into estuarine and marine food webs will increase. The massive expansion in aquaculture will also fuel exposure of humankind to contaminants, with farmed salmon, for example, already having been shown to have significantly higher concentrations of contaminants than wild populations.[2]

In addition, health effects can occur through psychological influences relating to socio-economic parameters. For example, degradation of natural systems detracts from human perception of beauty and enjoyment. Loss of amenities, ecology, produce and the potential for reductions in respective earnings, and especially tourism, can result in human stress with its associated health implications. Coupled with climate change and extreme weather events, this also creates uncertainty and undermines sustainability.

An example of loss of produce can be taken from the case of tributyltin (TBT). Tributyltin is an extremely toxic antifouling paint additive (although now banned) used to inhibit biological growth on ships' hulls. At extremely low concentrations, TBT can induce severe deformities in molluscs. It was estimated that in Arcachon Bay (France) alone, the use of TBT provoked a loss in revenue of 147 million US dollars through reduced oyster production.[3] Clearly there are also the issues of human consumption of the contaminated oysters!

Whilst, as will discussed in a subsequent section, the EU Water Framework Directive will strive to afford sustainability, current estimates suggest that, by 2015, UK estuaries and coastal water bodies at risk of failing to meet 'good' status will be 35% and 20%, respectively, for diffuse pollution, and 84% and 24%, respectively, for point source pollution.[4] This constitutes an interesting problem which should be analysed in more detail.

2 Public Perception

It is essential to consider perception of the public with respect to environmental issues as (with guidance from scientists) they eventually control those responsible for introducing the all important environmental legislation. The European Commission regularly requests and coordinates polls to gauge perception.[5] The latest results, albeit from 2007 (compared with data from 2004), are shown in Figure 1. It is of note that whilst climate change is at the top of the agenda for politicians, the public identify water pollution and man-made disasters (such as oil spills) on virtually an equivalent basis. Both relate to the current chapter. The same perception is apparent on the North American continent, particularly after the loss of amenities and the dramatic contamination associated with, for example, the Deepwater Horizon oil spill. Catastrophic events like this rekindle world recognition of environmental threats associated with pollution.

Other aspects that regularly achieve notoriety in the popular press relate to potential human exposures. Usually associated with drinking water (rather than saline), often emerging contaminants such as drugs (both prescription and illicit) are identified in drinking waters. It is interesting that many of these substances actually end up in sewage treatment works, with a proportion being discharged to the environment providing new potential routes for exposure. Emerging substances and nanoparticles are discussed later in this chapter.

As noted, where individuals' livelihoods are directly implicated with contamination/pollution issues, perspectives rapidly change and in the presence of litigation drivers, environmental issues receive a greater precedence whereby quantitative controls are of increasing relevance.

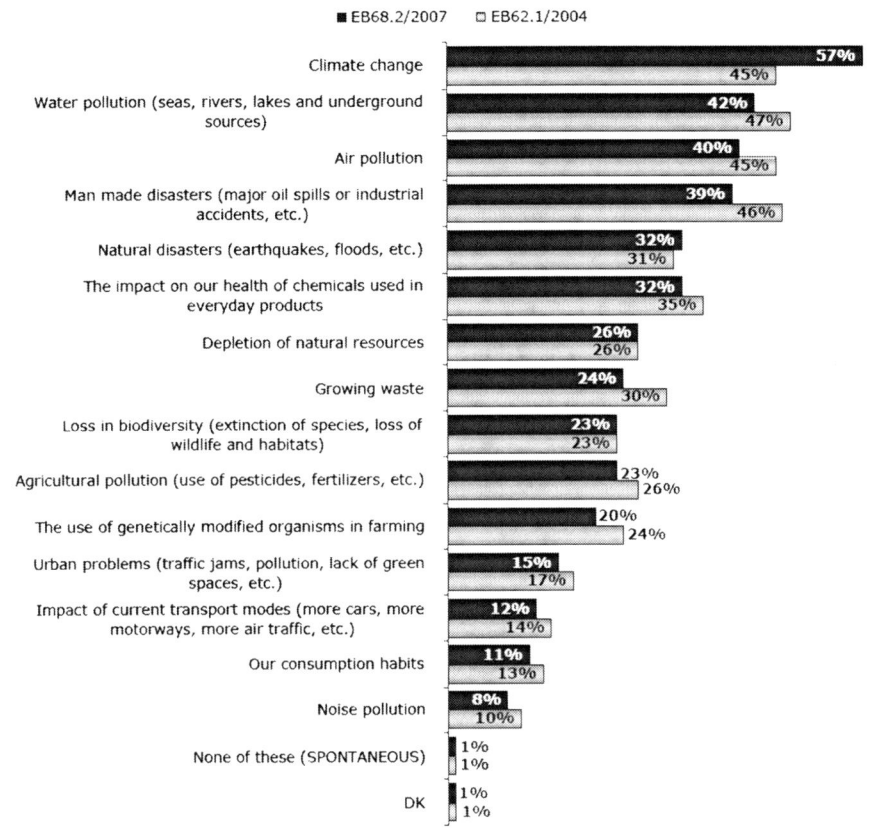

QF3 From the following list, please pick the five main environmental issues that you are worried about? (MAX. 5 ANSWERS) - % EU

■ EB68.2/2007   ▫ EB62.1/2004

Issue	EB68.2/2007	EB62.1/2004
Climate change	57%	45%
Water pollution (seas, rivers, lakes and underground sources)	42%	47%
Air pollution	40%	45%
Man made disasters (major oil spills or industrial accidents, etc.)	39%	46%
Natural disasters (earthquakes, floods, etc.)	32%	31%
The impact on our health of chemicals used in everyday products	32%	35%
Depletion of natural resources	26%	26%
Growing waste	24%	30%
Loss in biodiversity (extinction of species, loss of wildlife and habitats)	23%	23%
Agricultural pollution (use of pesticides, fertilizers, etc.)	23%	26%
The use of genetically modified organisms in farming	20%	24%
Urban problems (traffic jams, pollution, lack of green spaces, etc.)	15%	17%
Impact of current transport modes (more cars, more motorways, more air traffic, etc.)	12%	14%
Our consumption habits	11%	13%
Noise pollution	8%	10%
None of these (SPONTANEOUS)	1%	1%
DK	1%	1%

Figure 1 Attitudes of European citizens towards the environment: ranked environmental issues of concern.[5] The survey was requested by European Commission Directorate General Environment and was coordinated by Directorate General Communication. The report does not represent the point of view of the European Commission, interpretations and opinions contained in it are solely those of its authors. (© European Union, 1995–2011).

3 Priority Substances and Legislation

Whilst reviews on this topic are available[6–8] and the policy drivers are further discussed in Chapter 5, it is useful to update and summarise pertinent legislation and the theory behind selection of priority substances.

Generally, PBT (Persistence, Bioaccumulation and Toxicity) criteria are reviewed to select compounds of potential concern. Thus if a compound is toxic at low concentrations if it bioaccumulates in biota or it is persistent in the environment, it is deemed to be of potential concern. These parameters are tempered by other factors such as whether or not they are mutagenic, carcinogenic or disrupt endocrine systems. Another consideration is whether they are routinely identified in monitoring programmes.

Within the European Union, in the context of water policy, the substances listed in Table 1 have been selected as priority substances (according to Annex II of the Directive 2008/105/EC).[9,10] These are supplemented by eight 'certain other pollutants' which are regulated at the Community level.

Table 1 EU list of priority substances in the field of water policy.[10]

Name of priority substance	CAS number	EU number
Alachlor	15972-60-8	240-110-8
Anthracene	120-12-7	204-371-1
Atrazine	1912-24-9	217-617-8
Benzene	71-43-2	200-753-7
Brominated diphenylethers	n.a.	n.a.
Cadmium and its compounds	7440-43-9	231-152-8
C_{10-13}-chloroalkanes	85535-84-8	287-476-5
Chlorfenvinphos	470-90-6	207-432-0
Chlorpyrifos	2921-88-2	220-864-4
1,2-Dichloroethane	107-06-2	203-458-1
Dichloromethane	75-09-2	200-838-9
Di(2-ethylhexyl)phthalate (DEHP)	117-81-7	204-211-0
Diuron	330-54-1	206-354-4
Endosulfan	115-29-7	204-079-4
(α-endosulfan)	959-98-8	n.a.
Fluoranthene	206-44-0	205-912-4
Hexachlorobenzene	118-74-1	204-273-9
Hexachlorobutadiene	87-68-3	201-765-5
Hexachlorocyclohexane	608-73-1	210-158-9
(γ-isomer, Lindane)	58-89-9	200-401-2
Isoproturon	34123-59-6	251-835-4
Lead and its compounds	7439-92-1	231-100-4
Mercury and its compounds	7439-97-6	231-106-7
Naphthalene	91-20-3	202-049-5
Nickel and its compounds	7440-02-0	231-111-4
Nonylphenols	25154-52-3	246-672-0
(4-p-nonylphenol)	104-40-5	203-199-4
Octylphenols	1806-26-4	217-302-5
(p-tert-octylphenol)	140-66-9	n.a.
Pentachlorobenzene	608-93-5	210-172-5
Pentachlorophenol	87-86-5	201-778-6
Polyaromatic hydrocarbons	n.a.	n.a.
(Benzo[a]pyrene)	50-32-8	200-028-5
(Benzo[b]fluoroanthene)	205-99-2	205-911-9
(Benzo[g,h,i]perylene)	191-24-2	205-883-8
(Benzo[k]fluoroanthene)	207-08-9	205-916-6
(Indeno[1,2,3-c,d]pyrene)	193-39-5	205-893-2
Simazine	122-34-9	204-535-2
Tributyltin compounds	688-73-3	211-704-4
(Tributyltin cation)	36643-28-4	n.a.
Trichlorobenzenes	12002-48-1	234-413-4
(1,2,4-Trichlorobenzene)	120-82-1	204-428-0
Trichloromethane (Chloroform)	67-66-3	200-663-8
Trifluralin	1582-09-8	216-428-8

Specifically, these are carbon tetrachloride, DDT, cyclodiene pesticides, aldrin, dieldrin, endrin, isodrin and tetrachloroethylene.

Other 'emerging' contaminants and nanoparticles are discussed in subsequent sections and are likely to be supplemented by substances identified under the European REACH (Registration, Evaluation and Authorisation of Chemicals) programme. With a global production of chemicals of approximately 400 million tonnes and with about 100 000 different substances registered in the EU,[11] the REACH programme faces substantial challenges. Its basic aim is to protect human health and the environment and afford sustainability. To this end, all chemicals (with production exceeding 1 tonne) will need to be registered and evaluated with respect to hazardous properties, uses and exposure. Priority will be given to those substances with suspected dangerous properties and high exposure. The aim is to substitute the use of dangerous substances with less dangerous alternatives. One area that is proving controversial relates to nanoparticles (see section 5).

Water policy within Europe is set by the European Commission[12] primarily through the *Water Framework Directive 2000/60/EC* (WFD). The WFD aims to bring 'good status' in terms of ecological, chemical and physical quality by 2015, and covers rivers, lakes, coastal and transitional waters. It also streamlines EU legislation by incorporating previous Directives into the Framework. The driving force to improve the environmental quality is through the production and implementation of appropriate management plans and addresses both point and diffuse sources of pollution. From a chemical viewpoint, the priority pollutants specifically identified are those listed in Table 1. Whilst an admirable objective, as noted in section 1, estuaries and coastal water bodies at risk of failing to meet 'good' status in the UK alone are 35% and 20% for diffuse pollution and 84% and 24%, respectively, for point source pollution.[4]

More recently, in September 2010 the European Commission adopted a decision on the Marine Strategy Framework Directive that sets a 2020 target for good environmental status in terms of ecological diversity, health, productivity and sustainability.[13]

One group of compounds worthy of particular note is the Persistent Organic Pollutants (POPs). In 1962, Rachel Carson published a book[14] entitled *Silent Spring* that described how the pesticide DDT could biomagnify and contaminate food chains, harming animals, particularly at the higher trophic levels, including humankind. This book provoked public awareness of susceptibility from synthetic compounds. (It should, however, be noted that the US National Academy of Sciences has estimated that DDT prevented 500 million human deaths from malaria).[15]

In 2004, to protect human health from POPs, the United Nations Stockholm Convention became law. Following modifications introduced in May 2009, the Convention[16] requires that:

Parties must take measures to eliminate the production and use of the chemicals listed under Annex A.

Annex A:
- Aldrin
- Chlordane

- Chlordecone
- Dieldrin
- Endrin
- Heptachlor
- Hexabromobiphenyl
- Hexabromodiphenyl ether and heptabromodiphenyl ether
- Hexachlorobenzene (HCB)
- α-hexachlorocyclohexane
- β-hexachlorocyclohexane
- Lindane
- Mirex
- Pentachlorobenzene
- Polychlorinated biphenyls (PCB)
- Tetrabromodiphenyl ether and pentabromodiphenyl ether
- Toxaphene

Parties must take measures to restrict the production and use of the chemicals listed under Annex B.

Annex B:
- DDT
- Perfluorooctane sulfonic acid, its salts and perfluorooctane sulfonyl fluoride

Parties must take measures to reduce the unintentional releases of chemicals listed under Annex C.

Annex C:
- Polychlorinated dibenzo-p-dioxins (PCDD)
- Polychlorinated dibenzofurans (PCDF)
- Hexachlorobenzene (HCB)
- Pentachlorobenzene
- Polychlorinated biphenyls (PCB)

4 Emerging Contaminants

Whilst analytical techniques continue to improve, the number and frequency of detections of new chemicals increase; therefore, an attempt to cover the whole subject of emerging contaminants threatening the marine environment and human health would be almost impossible, and certainly incomplete. There are thousands of chemicals used in our modern lifestyles which have potential routes into the marine environment, through sewage treatment plant (STP) effluents, atmospheric inputs, coastal inputs and dispersion, accidental spills, *etc.* A definition for 'emerging pollutants' has been provided by the European Commission NORMAN research project[17] as "pollutants that are currently not included in routine monitoring programmes at the European level and which may be candidates for future regulation, depending on research on their (eco)toxicity, potential health effects and public perception, and on monitoring data regarding their occurrence in the various environmental compartments". Moreover, in environmental chemistry research, we normally find what we look for and, therefore, the number of chemicals of potential concern could be substantially underestimated. Consequently, this section only provides an overview, summarising the latest published research. It links different groups of currently emerging contaminants to proven contamination, and explores present and future potential sources. It is hoped that reading of this text will identify the main points of current concern on emerging contaminants and identify appropriate links to scientific publications for the compounds of interest.

Some of the so-called 'emerging contaminants' are known to have existed for decades. Political, economic, social, and scientific considerations have sometimes concealed them. However, novel and rapid technical advances in analytical chemistry and toxicological research, as well as an increasing social demand questioning whether or not our lifestyles affect our health, have influenced scientific reappraisals of these chemicals.

It is apparent that the chemical and pharmaceutical industries provide most chemicals and derivatives employed to produce modern commodities. They also supply other industries (*e.g.* the food industry) with what can be termed 'industrial chemicals'. 'Emerging contaminants', however, can be differentiated into two main groups as a function of their uses and looking at them from a human health perspective. The first group includes industrial chemicals (*e.g.* antifoaming agents, anticorrosive substances, *etc.*) where the individual consumer is not directly involved with its handling or use (*i.e.* passive exposure). In contrast, the second group includes emerging contaminants such as household chemicals (*e.g.* pharmaceuticals, fragrances, washing detergents, *etc.*) where individuals are directly in contact with the products (*i.e.* active exposure). The marine environment in almost all cases is the final repository for the majority of these anthropogenic (chemical) residues and metabolites. This potentially affords a threat to marine wildlife, ecosystem conservation and seafood production.

4.1 Industrial Emerging Contaminants

Industrial processes have released, to date, well-known contaminants that have been adopted as priority pollutants (see section 3). Other widely used industrial chemicals have included pesticides (*e.g.* benomyl, carbendazim, aldrin, endrin, ethion, malathion, biphenthrin, cypermethrin, diazinon, methoxychlor and dieldrin).[18] More recent examples of industrial chemicals are, for example, bisphenol A (BPA), which is being used as a primary raw material for the production of engineered plastics (*e.g.* polycarbonate/epoxy resins), food cans (lacquer coatings) and dental composites/sealants,[19] and pentachlorophenol, a fungicide that is widely used as a wood preservative (classified in 1999 by the International Agency for Research on Cancer as a possible human carcinogen), as well as other chlorophenols.[20,21]

Industrial contaminants and additives in food not only provide the potential for direct effects on humans, but are often recycled to the environment in sewage. Chemical food safety and environmental health issues have gained prominence due to the presence of natural toxins (*e.g.* shellfish toxins), processing-produced toxins (*e.g.* acrylamide, heterocyclic aromatic amines and furans), food allergens, heavy metals (*e.g.* lead, arsenic, mercury, cadmium), industrial chemicals (*e.g.* benzene and perchlorates), contaminants from packaging materials (*e.g.* phthalates), and 'unconventional' contaminants (*e.g.* melamine).[22] There is also great research interest, from a human health perspective, into the role of additives (*e.g.* dyes, preservatives/antioxidants, emulsifiers, *etc.*). Other potential organic contaminants include pesticides and dioxins or veterinary medicines (*e.g.* antibiotics). Coupled with priority pollutants, foods (including seafood and aquaculture products) afford a direct and important route of uptake of emerging contaminants by humans.

The chemical industry as the primary producer, has developed numerous chemicals to, for example, prevent the formation of foams at liquid interfaces in multiple industrial processes (*e.g.* sewage treatment plants, desalination plants, air-stripping processes, stirred fermenters, *etc.*) to reduce operational problems and hence costs.[23,24] Other industrial emerging contaminants are the complexing agents benzotriazole (BT) and tolyltriazole (TT) which are applied as anticorrosive agents (*e.g.* in cooling and hydraulic fluids, in antifreeze fluids, in aircraft as deicing fluids, in dishwashing liquids for silver protection), as antifogging agents, and as intermediates for the synthesis of various chemicals.[25] Benzotriazole and tolyltriazole can be considered as ubiquitous contaminants in the aquatic environment.[26] However, due to their physicochemical properties (*e.g.* low biodegradability and high hydrophobicity), they are not removed from sewage treatment plants and, coupled with their toxicity, pose a threat to estuarine and coastal environments. Other classes of complexing agents, such as sodium citrate, EDTA, sodium oxalate and tetrasodium pyrophosphate, have also been used in environmental applications (*e.g.* enhancing the desorption of pollutants from soil in contaminated land remediation).[27]

Phthalates, used primarily as plasticisers, are found everywhere in the environment (including foods). Their widespread use in so many products has led to calls that the compounds are probably under regulated. High levels of di-(2-ethylhexyl) phthalate (DEHP) and diisononyl phthalate (DINP) have been reported in shower and bath gels packed in soft PVC[28] and 2,6-di-isopropylnaphthalene (DIPN) and *n*-dibutylphthalate (DBP) were found in domestic and imported paper packages (less than 20% and more than 60%, respectively) in US marketplaces.[29] A major problem in determining these chemicals is that they are present in the materials which are typically employed within the analytical laboratories performing residue analyses and consequently blank controls are frequently contaminated. Phthalates in the environment have also been linked to endocrine disruption (see section 6).

Gasoline (petrol) additives include octane enhancers, antiknock compounds and oxygenates, as well as corrosion inhibitors, detergents and dyes. The final product also contains sulfur, nitrogen, oxygen and certain metals.[30] Fuel oxygenates, such as methyl *tert*-butyl ether (MTBE) and *tert*-amyl ethyl ether (TAEE) have been found in coastal environments.[31,32] These oxygenate compounds have largely replaced the use of organic lead compounds which polluted transport routes and the ecosystem in general (including humans) for decades. Indeed, anthropogenic lead can even be found in the deep ocean.[33]

Another emerging group of industrial chemicals are antioxidant substances, which are used in huge quantities in many areas of our modern life, including as food preservatives, fuel and oil stabilisers, to prevent oxidative degradation of plastics and in personal care products. For example, biodiesel and biodiesel blends are mainly vegetable oils or animal fats. These are composed of fatty acid monoalkyl esters (FAMEs) and can suffer oxidation during storage causing chemical degradation and the formation of products such as aldehydes, alcohols, and shorter chain carboxylic acids. These reduce fuel quality.[34] To prevent this, different types of natural and synthetic antioxidants are used for fuel and oil stabilisation. Among them are 2,6-di-*tert*-butyl-4-methylphenol, 2,2'-methylenebis(4-methyl-6-*tert*-butylphenol) and N,N'-di-2-butyl-1,4-phenylenediamine.

Generally, the main routes of chemicals to the marine environment are through surface water and sewage treatment plant discharges into receiving waters. However, other sources of industrial emerging contaminants into the environment are through indirect emissions, for example, in the case of the polybrominated diphenyl ether compounds (PBDEs) (see also section 3). These are industrial chemicals used as flame retardants in many polymeric materials such as furnishing foams, rigid plastics and textiles, therefore constituting a pathway of human passive exposure.[35] In addition, a secondary source of PBDEs emissions relates to e-waste storage facilities and landfills. These also act as a source of other priority and emerging contaminants, such as polybrominated biphenyls (PBBs), tetrabromobisphenol A (TBBPA), polybrominated phenols (PBPs), polychlorinated biphenyls (PCBs), polychlorinated dibenzo-*p*-dioxins and dibenzofurans (PCDD/Fs), polybrominated dibenzo-*p*-dioxins and dibenzofurans (PBDD/Fs) and chlorinated polycyclic aromatic hydrocarbons (Cl-PAHs).[36] Persistent organohalogen contaminants (*e.g.* PBDEs) are easily transported long distances and have been found accumulated in marine organisms from the Arctic.[37,38] Persistent organohalogen contaminants, in particular, are ubiquitous in all compartments (including water, sediment and biota from marine and estuarine environments.[39,40]

Perfluoroalkylated contaminants (PFAS) are industrial chemicals used for surface treatment in carpets and textiles, in polymer production (*e.g.* polytetrafluoroethylene, PTFE), in firefighting foams, *etc*. They, too, exhibit long-range atmospheric transport, and ubiquity.[41,42] The two most-studied PFAS compounds are perfluorooctane sulfonate (PFOS) and perfluorooctanoic acid (PFOA). As mentioned in section 3, they are included within the Stockholm Convention, and the United States Environmental Protection Agency (US EPA) considers both compounds to be carcinogenic. Studies of biota in the

Mediterranean Sea have shown that some fish contain extremely high concentrations of PFOA and PFOS,[43] endorsing concern about contamination of the marine environment.

Finally, a rather unusual class of emerging industrial substances are those which could be used as potential weapons in conflicts or in terrorism. With respect to aquatic environments, poisoning represents the largest threat, albeit primarily to drinking water supplies. Whilst organophosphorous compounds (*e.g.* Sarin, Soman, Tabun, V-gases, *etc.*) are considered priority chemical warfare agents (CWAs),[44–47] other more accessible compounds (*e.g.* the herbicide Paraquat, organophosphate insecticides, potassium cyanate) are candidates. In the marine environment there is also a legacy of contaminants such as CWAs (and explosives) remaining in, or being released during corrosion of sunken warships.

4.2 Other Emerging Contaminants

The second general group of emerging chemicals are, as mentioned earlier, those which can find a direct route of exposure to humans through lifestyle habits and necessities. Generically, these originate from household products containing toxins, such as paints, garden chemicals, pharmaceuticals, certain detergents, cleaners, personal care products, *etc.*[48]

Synthetic musks are within this group since these chemicals are used as the background product in many common fragrances, household cleaners, air fresheners, *etc*. These are mainly nitro-musk and polycyclic musk structures, such as 7-acetyl-1,1,3,4,4,6-hexamethyl-1,2,3,4-tetrahydronaphthalene (Tonalide; AHTN), 1,3,4,6,7,8-hexahydro-4,6,6,7,8-hexamethylcyclopenta-γ-2-benzopyran (Galaxolide;HHCB), 5-acetyl-1,1,2,6-tetramethyl-3-iso-propylindane (ATII), 4-acetyl-1,1-dimethyl-6-*tert*-butylindane (Celestolide; ADBI), 6-acetyl-1,1,2, 3,3,5-hexamethylindane (Phantolide; AHMI), 1-*tert*-butyl-3,5-dimethyl-2,4,6-trinitrobenzene (MX; musk xylene), and 4-*tert*-butyl-3,5-dinitro-2,6-dimethylacetophenone (MK; musk ketone).[49] Concerns have been expressed over their potential toxicity.[50] When these chemicals arrive at sewage treatment plants, they are only partially removed[51] and are released into receiving waters along with their byproducts (4-NH_2-musk xylene, 2-NH_2-musk xylene and 2-NH_2-musk ketone).[52] Compartmental distributions (dissolved, suspended particle associated and sedimentary) throughout an estuary, as well as in the coastal environment, have been reported.[53]

Classical detergent residues are another example of chemicals which have been studied globally for decades, such as linear alkylbenzene sulfonates (LAS), often in marine coastal sediments. More recently, another large pool of individual chemicals used as detergents, alkylphenols and alkylphenol ethoxylates (APEOs), have been studied due to their potential for causing estrogenic effects. Some are classified as endocrine disrupters (*e.g.* nonylphenol) and are therefore capable of interfering with the hormonal system of numerous organisms (see also section 7).[54–56]

Another very prominent group of emerging contaminants are the pharmaceuticals and personal care products (PPCPs) which comprise hundreds of chemicals. Clearly the pharmaceuticals are designed to be bioactive and include different families such as analgesics, antibiotics, antiepileptics, beta-blockers, blood-lipid regulators and contraceptives. These, combined with veterinary drugs, all have the potential to change physiological responses in aquatic organisms.[57,58] Personal care products frequently contain biocides, but also surfactants and fragrances, as discussed earlier in this section. Whilst they are generally less potent than drugs, they are used in massive quantities, representing probably a greater threat to the environment. Generally, information is still lacking with regard to the potential impact associated with the occurrence, fate and eco-toxicological effects of pharmaceuticals and personal care compounds in the environment, since relatively few compounds have been fully investigated. Residues of pharmaceutically active compounds (PhACs) and drugs of abuse (DAs) can therefore reach STPs in substantial amounts, often escaping degradation, and are then released into surface waters.[59] Environmental concentrations are low, but risks for the environment and human health cannot be excluded. Morphine, cocaine, methamphetamine and ecstasy all have potent pharmacological activities, and their presence as complex mixtures in surface waters may be toxic to aquatic organisms.[60] Moreover, recent studies suggest that pharmaceutical contaminants may pose long-term ecological risks through mixtures and the presence of their metabolites.[58] Psychiatric pharmaceuticals, such as anxiolytics, sedatives, hypnotics and antidepressants, are among the most prescribed active substances throughout the world. The occurrence of these widely used compounds has been reported in environmental matrices (wastewaters, surface, ground and drinking waters, soils, sediments, biosolids and tissues) and initial studies indicate that, for some, their high persistence and toxicity to non-target organisms justify concern.[61]

Sewage treatment plant effluents are often considered to be the most important point-sources of a wide range of emerging anthropogenic contaminants to aquatic systems. However, direct sources of UV filters (sunscreens), insect repellents and selected biocides could also be introduced directly into the marine environment from bathing areas and marinas.[62]

Finally, another group with an industrial origin but with direct human exposure are disinfection byproducts (DBPs). Whilst these have been managed and studied for the last 30 years, recent new concerns have arisen regarding adverse reproductive and developmental effects in human populations (through, for example, inhalation and dermal adsorption) as a result of exposure to DBPs.[63,64] Therefore, many drinking water utilities have changed their primary disinfectant from chlorine to alternatives (*e.g.* ozone, chlorine dioxide and chloramines), which generally reduce regulated trihalomethane and haloacetic acid levels, but can increase the levels of other potentially toxicologically important DBPs.[65] In desalination plants, the formation and speciation of DBPs is affected by the elevated concentrations of bromide and iodide in seawater and desalinated product water. When seawater or reverse osmosis permeate is chlorinated, bromoform ($CHBr_3$) and brominated haloacetic acids (*e.g.* monobromoacetic acid, dibromoacetic acid and bromochloroacetic acid) are found to be the prevalent DBP species. Less information is available on the formation of other halo-organic DBPs in desalination plants, such halo-acetonitriles (HANs), mutagen X compounds (MX), halonitromethanes (HNMs) and cyanogen bromides (CNBr), especially when desalinated waters are blended with organic-matter-rich source waters.[66]

It is clear that whilst known pollutants are regulated, horizon scanning is essential to protect the environment and human health from other potentially important novel contaminants. This topic is further discussed in section 7.

5 Nanoparticles

Nanotechnologies have been identified as one of eight key science and technology clusters that have the potential to affect wealth creation, change society and transform public services in the current decade (reviewed by Scown *et al.*).[67] The products of this rapidly expanding multibillion dollar industry stretch across many scientific disciplines, including chemistry, materials science, engineering, physics, biosciences, medicine and environmental sciences. A unifying feature of the products is their dimension of less than 100 nm (see Figure 2). At this size the materials concerned often have properties very different from those at a larger scale (termed bulk analogues). This is mainly due to a greater surface area : mass ratio that results in a greatly enhanced reactivity of the material. The very same special properties that make nanoparticles (NPs) useful are often also the properties that pose potential risks to humans and the environment under specific conditions. Very small alterations, for example slight changes in the particle size or attaching ligands to their surfaces, may alter their properties radically. These changing properties may subsequently lead to their toxicological properties changing unpredictably. The potential toxicological and long-term environmental properties of many NPs are not yet well understood. The major challenge is to fully understand the societal benefits of nanotechnology while identifying and minimising any adverse human and environmental impacts of NPs. Such understanding will come from an integrated multi-disciplinary strategy that connects environmental fate and behaviour with functional impacts at the various system levels. It is, therefore, essential that research, regulation and public dialogue recognise the diversity of the current and future applications. It is also important to be able to distinguish between free and fixed NPs, and between natural and manufactured NPs. Engineered NPs (ENPs) include accidentally produced (largely combustion-derived) and bulk-manufactured NPs (*e.g.* titanium dioxide, carbon black and alumina), all of which have been present for decades. We also need to consider the most recent generation of 'engineered' complex nanostructures with biomimetic properties that are the focus of more recent concerns.

5.1 Sources and Environmental Behaviour

Available databases on engineered nanoparticle (ENP) production are almost exclusively weight-based instead of a more suitable 'particle number-based' system, especially from a risk assessment (RA) perspective. The largest production levels are presently for carbon-based materials, including carbon black, *i.e.* fine elemental carbon demanded chiefly (>60% of worldwide production) by the tyre industry. This is projected to rise 4.3% per year through

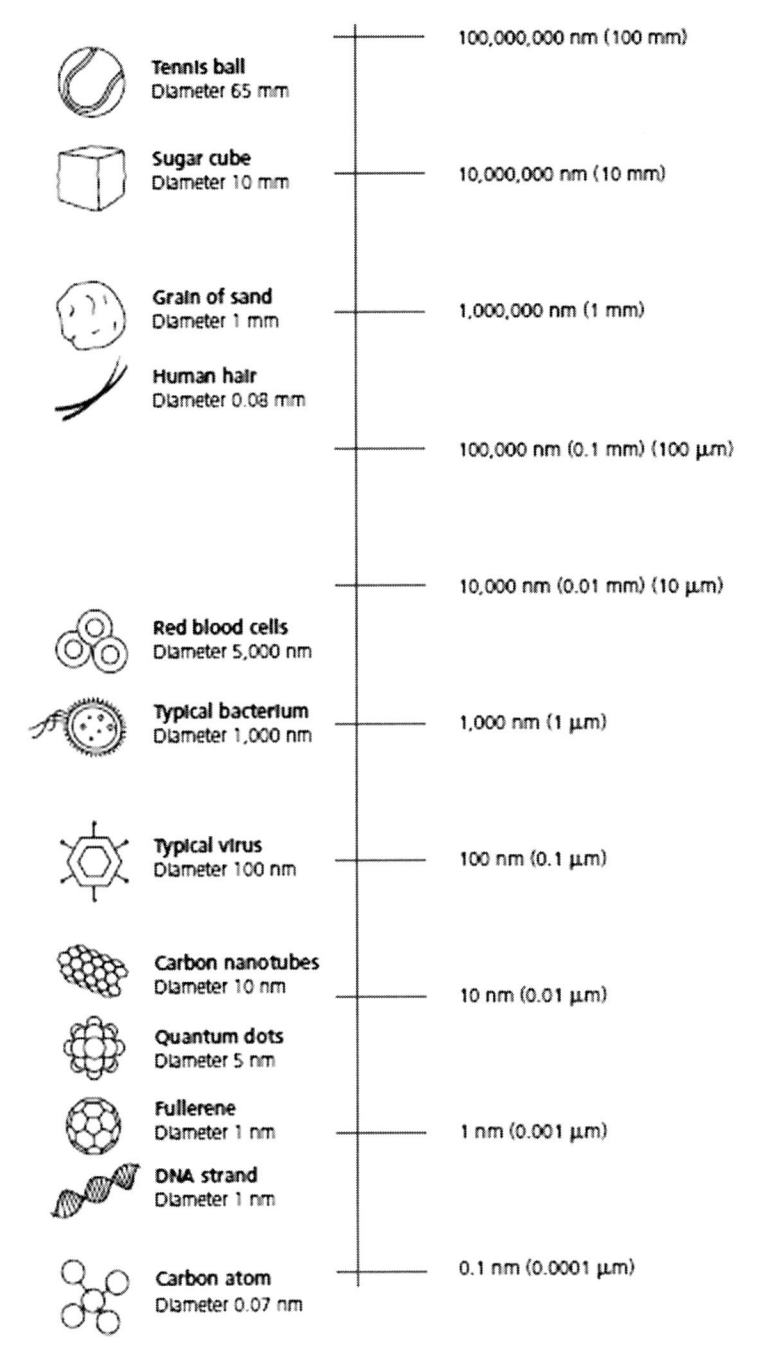

Tennis ball — Diameter 65 mm	100,000,000 nm (100 mm)
Sugar cube — Diameter 10 mm	10,000,000 nm (10 mm)
Grain of sand — Diameter 1 mm	1,000,000 nm (1 mm)
Human hair — Diameter 0.08 mm	
	100,000 nm (0.1 mm) (100 μm)
	10,000 nm (0.01 mm) (10 μm)
Red blood cells — Diameter 5,000 nm	
Typical bacterium — Diameter 1,000 nm	1,000 nm (1 μm)
Typical virus — Diameter 100 nm	100 nm (0.1 μm)
Carbon nanotubes — Diameter 10 nm	10 nm (0.01 μm)
Quantum dots — Diameter 5 nm	
Fullerene — Diameter 1 nm	1 nm (0.001 μm)
DNA strand — Diameter 1 nm	
Carbon atom — Diameter 0.07 nm	0.1 nm (0.0001 μm)

Figure 2 Length scale showing the nanometre in context. One nanometre (nm) is equal to one-billionth (1 000 000 000) of a metre, 10^{-9} m. Most structures of nanomaterials are between 1 and 100 nm in one or more dimensions. (Reproduced from a Royal Commission on Environmental Pollution report on *Novel Materials in the Environment: The Case of Nanotechnology*, © Crown Copyright 2008).

2013 to consumption levels of 11.6 million metric tonnes.[68] Other uses are: (1) reinforcement materials for packaging, fibres, mouldings, pipes and cables; pigmentation for inks and toners; UV protection or antistatic properties for coatings in marine, aerospace and industrial applications; (2) one-dimensional carbon, *i.e.* graphene, which is used in lubricants and more recently as a hydrogen store in fuel cells;[69] (3) carbon nanotubes that are the strongest materials known to man and thus are used as composite fibres and in brushes for electrical engines; and (4) carbon buckyballs applied as catalysts and in drug-delivery systems, optical devices and chemical sensors.[70]

Other significant commercial applications use metal-based NPs including: aluminium nanopowders for explosives;[71] iron in decontamination of soil and water polluted with chlorinated compounds and heavy metals;[72] cobalt in medical imaging;[73] silver as antimicrobial agents;[74] titanium (over 4 million tonnes per year) in paints and as pigments for substrate protection (reviewed by Sharma);[75] zinc oxide in coatings;[76] aluminium oxides in cosmetics; silicon dioxide (>1 million tonnes per year) as a bonding agent in paints (reviewed by Park);[77] cerium oxide as combustion catalyst;[77] and quantum dots which have most evident uses in biological research and medicine.[78]

Composites combine NPs with larger bulk-type materials as additives to enhance mechanical, thermal, flame retardant or other properties. These composites may also find their way into more complex nanostructures with biomimetic properties projected within the field of nanotherapeutics/targeted drug delivery.[79]

The nanoindustry is gradually becoming more concerned with the potential negative impacts it may impose on the natural world and has thus started supporting research into building suitable environmental fate and transport models for NPs.[80,81] There are a number of inherent 'nanoscience challenges' that are relevant for all environmental compartments and which hinder our understanding of the transport and fate of ENPs. First, the unknown baseline levels of NPs in air, soils or water because we have no reliable methods to distinguish between natural and manufactured NPs of similar structure and size; second, unidentified physico-chemical interactions between different ENPs or with organic and inorganic materials in the environment; third, incessant development of new functional groups and coatings influence how ENPs interact with each other and with other components in the environment, defying our established algorithms to model fate and transport predictions.

Atmospheric transport and fates of NPs are governed by diffusion, agglomeration, wet and dry deposition, and gravitational settling – processes that are relatively well understood for incidental ultrafine particles (mostly generated by combustion), and which may be applied for ENPs (reviewed by Harrison).[82] However, very little is known about the chemical processes involved, especially in case of NPs that are coated to prevent agglomeration. There is virtually nothing known about how aging, photodegradation or catalytic interactions influence the atmospheric chemistry of ENPs.

Terrestrial fate and mobility of ENPs are determined on the one hand by their physico-chemical properties and on the other hand by the type of soils.[83]

Predictions here are complicated by the presence of photosensitising natural materials in soils, such as humics and fulvics.

Aquatic ecosystems can receive ENPs from atmospheric deposition, leaching from soil and through direct inputs, such as wastewater discharges or from groundwater reservoirs.[84] Engineered nanoparticles can, therefore, potentially be transported long distances.[85] Given these diverse inputs, the aquatic environment is highly susceptible to contamination with certain ENPs. Fates of NPs in aqueous environments are controlled by solubility, dispersability, interactions with natural and anthropogenic chemicals, and biological and abiotic processes. These processes can have a major impact on the exposure of aquatic organisms to NPs, as cation-rich marine and estuarine environments favour aggregation and sedimentation, reducing the likelihood of transport within the water column.[86] Alterations in these conditions could also favour the stabilisation of NPs in the water column, enabling uptake and/or biotransformation by aquatic organisms and transport within water systems. General models for predicting this behaviour have yet to be developed.[86]

Due to the increasing release of NPs into the environment, either intentionally (*e.g.* by remediation of polluted land), or accidentally (*e.g.* from waste water sources), there is an urgent need to investigate the impacts on the environment, humans and other phyla. When considering environmental health, the adoption of a 'life cycle perspective' for industrial production and consumption is increasingly prevalent. In essence, a lifecycle perspective means an awareness of the material flows associated with a given product system, from the extraction of raw materials through the use of the product, to the disposal of wastes, as well as the fuels, electricity, chemicals and infrastructure required along the way. This formalised Life Cycle Assessment (LCA) methodology for quantifying the environmental impacts of products and services should also be applied for ENPs to include information on: (1) physical and chemical properties of the products, anticipated amounts, persistence, proposed routes of disposal and recycling; (2) hazard identification and characterisation: human and environmental exposures; (3) risk characterisation to see if there is a correlative link between the above characteristics of the various types of ENPs and those of conventionally used compounds, which may lead to reduction of data needs for risk assessment; and (4) recommendations on how the hazards could be eliminated and the risks reduced. In addition to the environmental and human health aspects, the related social, economic and ethical issues should be assessed using a cost-benefit framework.

Even though the scientific literature on nanotoxicology is growing exponentially,[87] no specific nanotoxicological mechanisms have been fully characterised to date and results are often contradictory or report on exposures to unrealistically elevated concentrations. Progress in understanding the effects of ENPs has been achieved largely through the use of surrogate particles with well-defined parameters under controlled laboratory conditions. The commonly assumed mechanism of nanotoxicity[88] associated with the nanoscale is the generation of reactive oxygen species (ROS) that causes oxidative stress and proinflammatory effects in a number of *in vivo* and *in vitro* models (reviewed in Scown et al.,[67] Landsiedal et al.[89] and Klaine et al.[90]). Following ROS induction, a series of biological responses are triggered by attack on DNA, proteins and lipid membranes. Although the role of NPs in ROS generation is not well understood, it appears to be related to the large particle surface area.

As mentioned in section 3, chemicals in Europe are presently regulated under the REACH Regulation which entered into force in 2007. The Commission has considered the issue of nanomaterials within REACH, and with the exception of certain specific types of novel nanomaterials (such as fullerenes), for most materials it considers that the size of the particle should not affect how it is classified, meaning that most nanomaterials will be classified and treated as if they were bulk scale materials. Furthermore, the REACH regulations only require that chemicals should be registered when production or importation levels exceed one tonne per annum. Whilst this is an appropriate trigger level for conventional chemicals, it can be questioned as to whether or not this is appropriate for all nanomaterials.

An overview of the regulatory gaps in the development, manufacture, supply, use and end-of-life of free ENPs across all current and future foreseeable applications of nanomaterials was carried out by the ESRC Centre for Business Relationships Accountability Sustainability and Society (BRASS). Their report emphasises that the majority of gaps arise largely from incomplete information about the implications of human and environmental exposure to NPs, rather than any major regulatory oversights.

In summary, nanotechnology has great potential to improve the quality of our lives through the production of novel materials and techniques. It can also help to improve the quality of our environment through, for example, remediation of polluted sites and reductions in fossil fuel combustion (with associated emissions). These potential benefits, however, need to be tempered with appropriate environmental risk assessments associated with nanoparticle release into the environment. Currently there are substantial gaps in our knowledge to drive these assessments and more scientific research is required to plug the gaps.

6 Plastics

Plastics comprise a wide range of synthetic or semi-synthetic organic polymers derived most frequently from petroleum-related products. They are produced in massive quantities (millions of tonnes annually) and are used in the manufacture of most everyday products.[91] Various chemicals are frequently added to alter properties (*e.g.* phthalate plasticisers) and to prolong longevity (*e.g.* by using antioxidants).

Whilst dealing, primarily, with chemical contaminants in this chapter, plastics represent an important global contaminant that warrants inclusion, even if their presence may not fit perfectly within the overall context.

It is only comparatively recently that the scale of marine contamination from plastics has been realised. Since plastics are light, strong, durable and inexpensive, their usage is massive. Coupled with the fact that they can persist for centuries[92] and are buoyant, it is perhaps not surprising that plastics make up between 60 and 80% of all marine debris.[92] Whilst we are all familiar with beaches strewn with washed up plastic debris (bottles, food containers, packaging, bags, rope fragments, netting, *etc.*), less conspicuous small plastic pellets and granules are ubiquitously distributed.[91] The sources of plastics can be from land-based discharges, atmospheric deposition and dumping from ships (although theoretically Annex V of the MARPOL convention restricts at-sea discharge of garbage and bans sea disposal of plastics).[93]

Biological effects from large-scale plastics generally relate to entanglement or ingestion. Entanglement with plastic debris (and especially discarded fishing gear) impedes agility and constricts growth, impairing survival. Effects associated with smaller particles and microplastics are, from a contaminant perspective, more appropriate for discussion in this chapter. The global distribution of microplastics (<5 mm particles) is a comparatively recent discovery. Owing to their buoyancy, they travel primarily in the upper pelagic layer of the oceans, affording a ubiquitous distribution.[91,94] Their sources include degradation of larger pieces of plastic; raw manufacturing materials and from their use as abrasives in cleaning products.

Through their properties, microplastics are carried in the sea to gyres where they can accumulate. For example, in the North Pacific Central gyre, the abundance and mass of neustonic plastic has been reported as 334 271 pieces km^{-2} and 5 114 g km^{-2}, respectively.[95] This compares with a plankton abundance approximately five times higher but a mass of plastic approximately six times higher than that of the plankton.[95] Mechanisms of abrasion and degradation of the plastics are poorly understood.[96] Polymeric materials found in marine environments include acrylic, alkyd, poly(ethylene–propylene), polyamide (nylon), polyester, polyethylene, polymethylacrylate, polypropylene and polyvinyl alcohol.[91] Whilst they can, on degradation, release endocrine-disrupting phthalates and toxic antioxidants, the major concern relating to the plastics relates to their ability to sorb and concentrate hydrophobic organic pollutants and act as a carrier for them.[97,98] Indeed, this has led to their use in global pollution monitoring of POPs through the International Pellet Watch project.[99,100]

7 Complex Mixtures: Causality of Effects

The priority pollutants described in section 3 oversimplify the context of contamination/pollution. In essence, pollutants rarely occur as single compounds in isolation. Rather, they are usually present as highly complex mixtures affording multiple exposures to the biota. This renders toxicological implications far more complicated. For example, sewage outfalls[101] and landfill leachates[48,102] contain cocktails of priority pollutants and many of the emerging contaminants discussed in section 4 (including pharmaceuticals, personal care products, surfactants, fragrances, biocides, *etc.*). Combine this situation with elevated levels of nutrients, ammonia and particulates, and evaluation of the myriad of potential interactions contributing to toxicity is rendered extremely difficult. It is further complicated by the presence of carcinogens (such as

PAH)[101] and endocrine disrupters[103,104] [including oestrone, 17β-oestradiol and 17α-ethinylestradiol (the contraceptive pill)].[105] Couple in factors associated with bioavailability (such as particle occlusion associated with, for example, PAH in road run-off particles),[101,106,107] and implications of acute *versus* chronic exposure mechanisms, assessment and apportionment of effects is a substantial challenge.

So, how do we overcome these problems? Exposure pathways, bioavailability and toxic potential can be measured effectively by investigating the organisms themselves. If bioavailability is in question, biological fluids and tissues can be analysed to assess pollutant uptake by biota.[108,109] In addition, drivers to assess pollution are directed to toxicological potential. If the contaminants present do not invoke damage to organisms (including the human consumption of seafood) then problems could be considered to be of minor potential. However, although toxicological measures frequently relate to lethality, much more subtle impairment can occur at very low concentrations. To allow for this, 'biomarkers' of response can be invoked to detect very subtle responses that impair health. These can very usefully be incorporated within monitoring strategies (together with chemical assessments) to provide a more rounded appraisal of ecological change in response to contamination.[106] The WFD encourages this more complete approach.

Alternatively, another approach is to use a technique called 'Toxicity Identification and Evaluation' (TIE). This was initially introduced by the US Environmental Protection Agency.[110] Basically, TIE uses environmental sample extracts that are fractionated to investigate which substances within the environment actually effect toxicological responses (measured through bioassays). The fractionations can be increasingly refined to identify the causative toxic agents. Various analytical separations can be used to fractionate, although typically high-performance liquid chromatography (HPLC) is chosen. Fractions provoking toxic effects (identified through the bioassays), are then fully characterised by quantitative high-resolution gas chromatography-mass spectrometry or liquid chromatography-mass spectrometry. Using this approach with a bioassay yeast (that had been transfected with the human oestrogen receptor ERα gene linked to a reporter gene system), endocrine disruption from a sewage outfall could be linked with 17β oestradiol.[105] In a more general assessment of acute toxicity in UK estuaries, Thomas *et al.*[111,112] and Matthiessen and Law[113] identified the compounds that specifically caused toxic effects. These are summarised in Table 2. It is interesting that they are estuary specific and do not adhere to conventional priority pollutant associations.

8 Climate Change and Pollutants

Impacts of climate change on the marine environment are manifold. In the context of pollution, however, it is useful to explore potential interactions that can impact ecosystems, resources and services, and consequently humankind.

Whilst pollution can act to deleteriously affect biota, it rarely impacts in isolation and other stresses such as salinity, temperature, hypoxia and UV can often contribute (particularly in transitional waters). Even the presence and susceptibility to disease/parasites can change. Whilst these multiple stressors

Table 2 Substances present in UK estuarine waters which have been tentatively identified by TIE as contributing to toxic effects (as assessed using the copepod *Tisbe battagliai*). (Data summarised by Matthiessen and Law[113] and based on Thomas *et al.*)[111,112] Note: Other contributary substances remain to be identified.

Estuary	Substance
Tyne	Chlorobenzene acetonitrile
	4-Chloro-3,5-xylenol
	Methylacridine
	Methylpyridine amine
	Monomethyl-trimethyl fluorenes
	Monomethyl-trimethyl naphthalenes
	Naphthalene amine
	Nonylphenol
	Pentachlorophenol
	Trichlorophenol
	Tetrachlorophenol
	Triphenylphosphine sulphide
Tees	Atrazine
	Carbophenothion methylsulphoxide
	4-Chloro-3,5-dimethylphenol
	Diethylnaphthalene carboxamide
	Dimethylbenzoquinone
	Dimethylnaphthalene carboxamide
	Nonylphenol
Mersey	Dieldrin
	Dodecylphenol
Milford Haven	Bis(2-ethylhexyl)phthalate

are omnipresent, climate change can exacerbate them through the processes shown in Figure 3. These are discussed in more depth by Schiedek *et al.*[114]

Direct impacts of climate change on biota and ecosystems occur through alterations in physico-chemical parameters such as temperature, salinity, turbidity, pH, *etc*. This can alter niche locations to consequently exceed tolerances typically encountered by the biota. Changes in precipitation, and hence freshwater run-off (and increasingly frequent extreme events) also produce changes affording the potential for remobilisation of contaminants.

Changes in environmental conditions can also directly alter sources and hence composition of contaminants. For example, choices of different, more appropriate and profitable crops, and the selection of biocide regimes to protect them will alter, as will precautionary measures to protect human health. In addition, changes to transport, partitioning, biological uptake, metabolism and subsequent biological effects, with trophic implications, will be affected.

An example that is particularly pertinent relates to the plight of the eel population. It has decreased, for example, in London's River Thames by 98% in just five years.[115] Why? This might be because of climate change, pollutants, new engineering structures, parasites, migratory behaviour *etc.*, or a combination of any or all of them. Nobody knows.

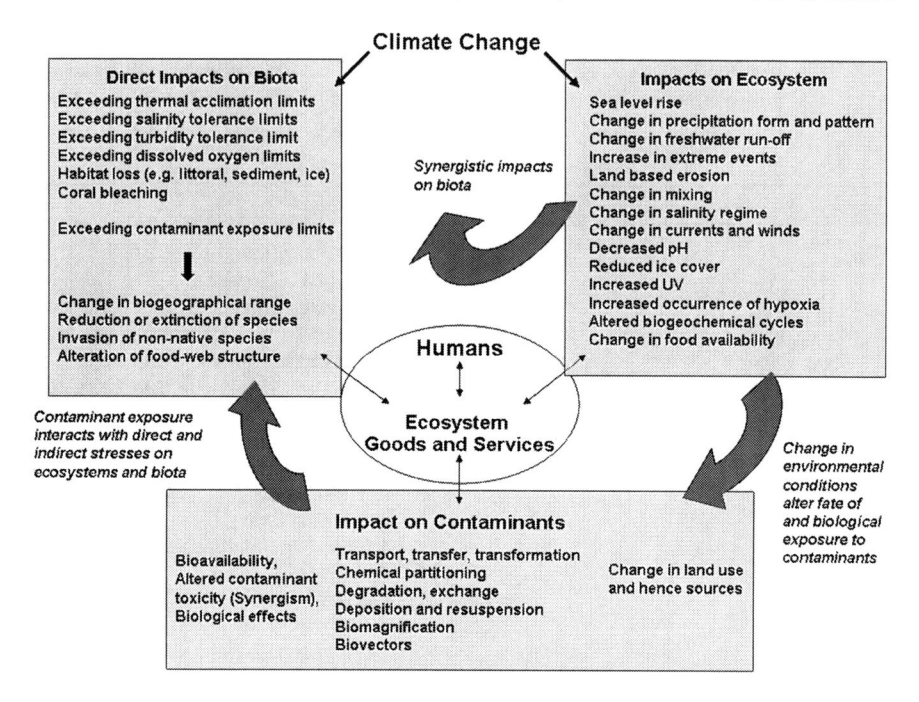

Figure 3 Overview of climate change impacts on ecosystems and biota, and how they interact with contaminants, and their fate and effects. (Taken from Schiedek et al.).[114]

9 Future Issues

Human health implications relating to chemical pollution of estuarine and marine systems, with few exceptions, tend to relate to longer-term (chronic) exposure generally through consumption of contaminated seafood and socio-economic/health implications. The latter are associated with degradation of the environment with losses of amenities, services and produce. Exceptions are possibly most notable in the case of incidents such as the Deepwater Horizon oil spill in the USA and shipping accidents such as the MSC Napoli in the UK, where coastal ecosystems are rapidly impacted.

With respect to chronic contamination and environmental degradation, certain topics are identified where our comprehension of ecosystem functioning is lacking. These, in our opinion, include: (1) the combinations of stresses (both natural and anthropogenic, including climate change), especially in transitional and coastal waters; (2) our understanding of interactive effects associated with complex mixtures of contaminants and identification of the causative agents of damage; (3) comprehension of exposure pathways, uptake and bioavailabilty; and (4) the vulnerability of ecosystem components, including early life stages of organisms.

As noted in sections 4 and 5, humankind is constantly changing products that are emitted as 'wastes' into the environment. It is important to evaluate 'emerging' substances (including nanoparticles) and to appropriately adjust impact assessment practices so that they are suitable to evaluate risks to ecosystem amenities, services, produce and health.

Clearly, it is necessary to provide research which can guide policy drivers to afford sustainability of our ecosystems and address socio-economic and health implications. Emerging issues require constant attention, and novel evaluation techniques are required to guide the policy drivers.

References

1. P. T. C. Harrison, in *Pollution, Causes, Effects and Control*, ed. R. M. Harrison, The Royal Society of Chemistry, 4th edn., 2001.
2. R. A. Hites, J. A. Foran, D. O. Carpenter, M. C. Hamilton, B. A. Knuth and S. J. Schwager, *Science*, 2004, **303**(5655), 226.
3. C. Alzieu, *Mar. Environ. Res.*, 1991, **32**, 7.
4. Environment Agency, *The Unseen Threat to Water Quality: Diffuse Water Pollution in England and Wales Report*, May 2007, http://www.environment-agency.gov.uk/static/documents/Research/geho0207bzlvee_1773088.pdf (last accessed 08/02/2011).
5. European Commission, *Attitudes of European Citizens towards the Environment*, Special Eurobarometer 295/Wave 68.2, 2008, http://ec.europa.eu/public_opinion/archives/ebs/ebs_295_en.pdf (last accessed 08/02/2011).
6. B. Crathorne, Y. J. Rees and S. France, in *Pollution, Causes, Effects and Control*, ed. R. M. Harrison, The Royal Society of Chemistry, 4th edn., 2001.
7. R. Macrory, in *Pollution, Causes, Effects and Control*, ed. R. M. Harrison, The Royal Society of Chemistry, 4th edn., 2001.
8. J. Kinniburgh, in *An Introduction to Pollution Science*, ed. R. M. Harrison, The Royal Society of Chemistry, 2006.
9. European Commission, *Annex II of the Directive 2008/105/EC*, 2008, http://eur-lex.europa.eu/LexUriServ/LexUriServ.do?uri=OJ:L:2008:348:0084:0097:EN:PDF (last accessed 08/02/2011).
10. European Commission, *Priority Substances and Certain Other Pollutants*, 2008. http://ec.europa.eu/environment/water/water-framework/priority_substances.htm (last accessed 08/02/2011).
11. European Commission, *White Paper – Strategy for a Future Chemicals Policy*, Commission of the European Communities COM (2001)88 final, 27.2.2001.
12. European Commission, *Water is for Life: How the Water Framework Directive helps Safeguard Europe's Resources*, 2010, pp. 25, http://ec.europa.eu/environment/water/pdf/WFD_brochure_en.pdf accessed 08/02/2011 (last accessed 08/02/2011).
13. European Commission, *A Marine Strategy Directive to save Europe's Seas and Oceans.*, 2010, http://ec.europa.eu/environment/water/marine/index_en.htm (last accessed 08/02/2011).

14. R. Carson, *Silent Spring*, Houghton Mifflin Co., Boston, USA, 1962.
15. National Academy of Sciences, *The Life Sciences*, National Academy of Sciences Press, Washington DC, USA, 1970.
16. *Stockholm Convention on Persistent Organic Pollutants*, 2010, http://chm.pops.int/Convention/tabid/673/language/en-GB/Default.aspx (last accessed 08/02/2011).
17. European Commission, *NORMAN Research Project*, http://www.norman-network.net (last accessed 08/02/2011).
18. K. E. Murray, S. M. Thomas and A. A Bodour, *Environ. Pollut.*, 2010, **158**, 3462.
19. W. T. Tsai, *J. Environ. Sci. Health, Part C*, 2006, **24**, 225.
20. G. S. Cooper and S. Jones, *Environ. Health Perspect.*, 2008, **116**, 1001.
21. W. E. Pepelko, D. W. Gaylor and D. Mukerjee, *Toxicol. Ind. Health*, 2005, **21**, 93.
22. L. S. Jackson, *J. Agric. Food Chem.*, 2009, **57**, 8161.
23. Y. L. Sheng and C. H. Clatterbuck, *Soc. Adv. Mater. Process Eng. Quart.*, 1991, **22**, 52.
24. R. S. Bhute, *J. Sci. Ind. Res.*, 1971, **30**, 241.
25. D. Voutsa, P. Hartmann, C. Schaffner and W. Giger, *Environ. Sci. Pollut. Res.*, 2006, **13**, 333.
26. W. Giger, C. Schaffner and H. P. E. Kohler, *Environ. Sci. Technol.*, 2006, **40**, 7186.
27. K. Subramaniam, C. Stepp, J. J. Pignatello, B. Smets and D. Grasso, *Environ. Eng. Sci.*, 2004, **21**, 515.
28. J. P. Amberg-Muller, U. Hauri, U. Schlegel, C. Hohl and B. J. Bruschweiler, *J. Consumer Protect. Food Safety*, 2010, **5**, 429.
29. K. Zhang, G. O. Noonan and T. H. Begley, *Food Addit. Contam., Part A*, 2008, **25**, 1416.
30. K. Demirbas and A. Sahin-Demirbas, *Energy Sources, Part B*, 2010, **5**, 243.
31. R. da Silva, E. W. de Menezes and R. Cataluna, *Quim. Nova*, 2008, **31**, 980.
32. C. Guitart, J. M. Bayona and J. W. Readman, *Chemosphere*, 2004, **57**, 429.
33. T. O. Richter, H. C. de Stigter, W. Boer, C. C. Jesus and T. C. E. van Weering, *Deep-Sea Res., Part I*, 2009, **56**, 267.
34. S. Jain and M. P. Sharma, *Renewable Sustainable Energy Rev.*, 2010, **14**, 667.
35. S. Harrad, C. A. de Wit, M. A. E. Abdallah, C. Bergh, J. A. Bjorklund, A. Covaci, P. O. Darnerud, J. de Boer, M. Diamond, S. Huber, P. Leonards, M. Mandalakis, C. Oestman, L. S. Haug, C. Thomsen and T. F. Webster, *Environ. Sci. Technol.*, 2010, **44**, 3221.
36. H. G. Ni, H. Zeng, S. Tao and E. Y. Zeng, *Environ. Toxicol. Chem.*, 2010, **29**, 1237.
37. B. C. Kelly, M. G. Ikonomou, J. D. Blair and F. Gobas, *Environ. Sci. Technol.*, 2008, **42**, 7069.
38. R. J. Letcher, J. O. Bustnes, R. Dietz, B. M. Jenssen, E. H. Jørgensen, C. Sonne, J. Verreault, M. M. Vijayan and G. W. Gabrielsen, *Sci. Total Environ.*, 2010, **408**, 2995.

39. G. T. Yogui and J. L. Sericano, *Environ. Int.*, 2009, **35**, 655.
40. S. Tanabe, K. Ramu, T. Isobe and S. Takahashi, *J. Environ. Monit.*, 2008, **10**, 188.
41. A. Jahnke, S. Huberc, C. Ternme, H. Kylin and U. Berger, *J. Chromatogr., A*, 2007a, **1164**, 1.
42. A. Jahnke, U. Berger, R. Ebinghaus and C. Temme, *Environ. Sci. Technol.*, 2007b, **41**, 3055.
43. V. Nania, G. E. Pellegrini, L. Fabrizi, G. Sesta, P. de Sanctis, D. Lucchetti, M. di Pasquale and E. Coni, *Food Chem.*, 2009, **115**, 951.
44. N. V. Kokshareva, M. G. Prodanchuk, V. F. Tkach and M. L. Zinovieva, *Medical Treatment of Intoxications and Decontamination of Chemical Agents in the Area of Terrorist Attack*, ed. C. Dishovsky, A. Pivovarov and H. Benschop, Springer, Dordrecht, The Netherlands, 2006, 11, p. 101.
45. O. J. Mika and I. Masek, *Chem. Listy*, 2008, **102**, 255.
46. S. Le Moullec, A. Begos, V. Pichon and B. Bellier, *J. Chromatogr., A*, 2006, **1108**, 7.
47. P. Z. Ruckart and M. Fay, *J. Environ. Health*, 2006, **69**, 9.
48. R. J. Slack, J. Gronow and N. Voulvoulis, *Sci. Total Environ.*, 2005, **337**, 119.
49. W. D. Wombacher and K. C. Hornbuckle, *J. Environ. Eng. (ASCE)*, 2009, **135**, 1192.
50. T. Luckenbach and D. Epel, *Environ. Health Perspect.*, 2005, **113**, 17.
51. D. Herren and J. D. Berset, *Chemosphere*, 2000, **40**, 565.
52. G. G. Rimkus, R. Gatermann and H. Huhnerfuss, *Toxicol. Lett.*, 1999, **111**, 5.
53. N. R. Sumner, C. Guitart, G. Fuentes and J. W. Readman, *Environ. Pollut.*, 2010, **158**, 215.
54. S. A. Snyder, P. Westerhoff, Y. Yoon and D. L. Sedlak, *Environ. Eng. Sci.*, 2003, **20**, 449.
55. A. Soares, B. Guieysse, B. Jefferson, E. Cartmell and J. N. Lester, *Environ. Int.*, 2008, **34**, 1033.
56. K. Fent, A. A. Weston and D. Caminada, *Aquatic Toxicol.*, 2006, **76**, 122.
57. F. A. Caliman and M. Gavrilescu, *CLEAN – Soil, Air, Water*, 2009, **37**, 277.
58. M. D. Celiz, J. Tso and D. S. Aga, *Environ. Toxicol. Chem.*, 2009, **28**, 2473.
59. M. Petrovic, M. J. L. de Alda, S. Diaz-Cruz, C. Postigo, J. Radjenovic, M. Gros and D. Barcelo, *Philos. Trans. R. Soc. London, Ser. A*, 2009, **367**, 3979.
60. E. Zuccato and S. Castiglioni, *Philos. Trans. R. Soc. London, Ser. A*, 2009, **367**, 3979.
61. V. Calisto and V. I. Esteves, *Chemosphere*, 2009, **77**, 1257.
62. K. H. Langford and K. V. Thomas, *J. Environ. Monit.*, 2008, **10**, 894.
63. C. Legay, M. J. Rodriguez, J. B. Serodes and P. Levallois, *Sci. Total Environ.*, 2010, **408**, 456.
64. S. Chowdhury, P. Champagne and P. J. McLellan, *J. Environ. Manag.*, 2009, **90**, 1680.
65. S. D. Richardson, *TrAC, Trends Anal. Chem.*, 2003, **22**, 666.

66. E. Agus, N. Voutchkov and D. L. Sedlak, *Desalination*, 2009, **237**, 214.
67. T. M. Scown, R. van Aerle and C. R. Tyler, *Crit. Rev. Toxicol.*, 2010, **40**, 653.
68. World Carbon Black Industry Study with Forecasts for 2013 & 2018, http://www.freedoniagroup.com/ (last accessed 09/02/2011).
69. S. Patchknovskii, J. S. Tse, S. N. Yurchenko, L. Zhechkov, T. Heine and G. Seifert, *Proc. Natl. Acad. Sci. U. S. A.*, 2005, **102**, 10439.
70. J. Kong, N. R. Franklin, C. Zhou, M. G. Chapline, S. Peng, K. Cho and H. Dai, *Science*, 2000, **287**, 622.
71. D. Spitzer, M. Comet, C. Baras, V. Pichot and N. Piazzon, *J. Phys. Chem. Solids*, 2010, **71**, 100.
72. B. Karn, T. Kuiken and M. Otto, *Environ. Health Persp.*, 2009, **117**, 1823.
73. V. F. Puntes, K. M. Krishnan and A. P. Alivisatos, *Science*, 2001, **291**, 2115.
74. X. L. Cao, C. Cheng, Y. L. Ma and C. S. Zhao, *J. Mater. Sci. Mater. Med.*, 2010, **21**, 2861.
75. V. K. Sharma, *J. Environ. Sci. Health, Part A*, 2009, **44**, 1485.
76. F. Alvi, M. K. Ram, H. Gomez, R. K. Joshi and A. Kumar, *Polym. J.*, 2010, **42**, 935.
77. B. Park, in *Nanotechnology Consequences for Human Health & the Environment*, ed. R. E. Hester and R. M. Harrison, The Royal Society of Chemistry, Cambridge, UK, *Issues in Environ. Sci. Technol.*, 2009, **24**, 7.
78. A. P. Alivisatos, W. W. Gu and C. Larabell, *Ann. Rev. Biomed. Eng.*, 2005, **7**, 55.
79. M. Guillot-Nieckowski, S. Eisler and F. Diederich, *New J. Chem.*, 2007, **31**, 1111.
80. A. B. A. Boxall, Q. Chaudhry, C. Sinclair, A. Jones, R. Aitken, B. Jefferson and C. Watts, *Report by Central Science Laboratory for Department for Environment, Food and Rural Affairs,* Her Majesty's Government, 2008, UK.
81. N. C. Mueller and N. Nowack, *Environ. Sci. Technol.*, 2008, **42**, 4447.
82. R. Harrison, in *Nanotechnology Consequences for Human Health & the Environment*, ed. R. E. Hester and R. M. Harrison, The Royal Society of Chemistry, Cambridge, UK, *Issues in Environ. Sci. Technol.*, 2009, **24**, 35.
83. B. K. G. Theng and G. Yuan, *Elements*, 2008, **4**, 395.
84. M. F. Schaller and Y. Fan, *J. Geophys. Res. Atmos.*, 2009, **114**, 22.
85. B. S. Caruso and H. E. Dawson, *Environ. Monit. Assess.*, 2009, **153**, 405.
86. M. Baalousha, A. Manciulea, S. Cumberland, K. Kendall and J. R. Lead, *Environ. Toxicol. Chem.*, 2008, **27**, 1875.
87. EPA 100/B-07/001 Nanotechnology White Paper, 2007, http://www.epa.gov/osa/pdfs/nanotech/epa-nanotechnology-whitepaper-0207.pdf (last accessed 10/02/2011).
88. K. Donaldson, P. H. Beswick and P. S. Gilmour, *Toxicol. Lett.*, 1996, **88**, 293.
89. R. Landsiedel, M. D. Kapp, M. Schulz, K. Wiench and F. Oesch, *Mutation Res.*, 2009, **681**, 241.

90. S. J. Klaine, P. J. J. Alvarez, G. E. Batley, T. F. Fernandes, R. D. Handy, D. Y. Lyon, S. Mahendra, M. J. Mclaughlin and J. R. Lead, *Environ. Toxicol. Chem.*, 2008, **27**, 1825.
91. R. C. Thompson, Y. Olsen, R. P. Mitchell, A. Davis, S. J. Rowland, A. W. G. John, D. McGonigle and A. E. Russell, *Science*, 2004, **304**, 838.
92. J. G. B. Derraik, *Mar. Pollut. Bull.*, 2002, **44**, 842.
93. J. B. Pearce, *Mar. Pollut. Bull.*, 1992, **24**, 586.
94. S. Moret-Ferguson, K. L. Law, G. Proskurowski, E. K. Murphy, E. E. Peacock and C. M. Reddy, *Mar. Pollut. Bull.*, 2010, **60**, 1873.
95. C. J. Moore, S. L. Moore, M. K. Leecaster and S. B. Weisberg, *Mar. Pollut. Bull.*, 2001, **42**(12), 1297.
96. P. L. Corcoran, M. C. Biesinger and M. Grifi, *Mar. Pollut. Bull.*, 2009, **58**, 80.
97. E. L. Teuten, S. J. Rowland, T. S. Galloway and R. C. Thompson, *Environ. Sci. Technol.*, 2007, **41**, 7759.
98. M. A. Browne, T. Galloway and R. Thompson, *Integr. Environ. Assess. Manage.*, 2007, **3**, 559.
99. H. Takada, *Mar. Pollut. Bull.*, 2006, **52**, 1547.
100. Y. Ogata, H. Takada, K. Mizukawa, H. Hirai, S. Iwasa, S. Endo, Y. Mato, M. Saha, K. Okuda, A. Nakashima, M. Murakami, N. Zurcher, R. Booyatumanondo, M. P. Zakaria, L. Q. Dung, M. Gordon, C. Miguez, S. Suzuki, C. Moore, H. K. Karapanagioti, S. Weerts, T. McClurg, E. Burresm, W. Smith, M. Van Velkenburg, J. Selby Lang, R. C. Lang, D. Laursen, B. Danner, N. Stewardson and R. C. Thompson, *Mar. Pollut. Bull.*, 2009, **58**, 1437.
101. J. Readman, in *An Introduction to Pollution Science*, ed. R. M. Harrison, The Royal Society of Chemistry, Cambridge, UK, 2006.
102. T. Eggen, M. Moeder and A. Arukwe, *Sci. Total Environ.*, 2010, **408**, 5147.
103. T. Colborn, D. Dumanoski and J. Peterson Myers, *Our Stolen Future*, http://www.ourstolenfuture.org/Basics/chemlist.htm (last accessed 09/02/2011).
104. European Commission, *Staff Working Document on the Implementation of the "Community Strategy for Endocrine Disrupters" – A Range of Substances Suspected of Interfering with the Hormone Systems of Humans and Wildlife*, (COM (1999) 706), (COM (2001) 262) and (SEC (2004) 1372), SEC(2007) 1635, 2007, http://ec.europa.eu/environment/endocrine/documents/sec_2007_1635_en.pdf (last accessed 09/02/2011).
105. K. V. Thomas, M. Hurst, P. Matthiessen and M. J. Waldock, *Environ. Toxicol. Chem.*, 2001, **20**, 2165.
106. M. N. Moore, M. H. Depledge, J. W. Readman and D. R. Paul Leonard, *Mutation Res.*, 2004, **552**, 247.
107. J. W. Readman, R. F. C. Mantoura and M. M. Rhead, *Fresenius Z. Anal. Chim.*, 1984, **319**, 126.
108. G. Fillmann, G. M. Watson, E. Francioni, J. W. Readman and M. H. Depledge, *Mar. Environ. Res.*, 2002, **54**, 823.

109. G. Fillmann, G. M. Watson, M. Howsam, E. Francioni, M. H. Depledge and J. W. Readman, *Environ. Sci. Technol.*, 2004, **38**, 2649.
110. D. I. Mount and D. M. Anderson-Carnahan, *Methods for Aquatic Toxicity Identification Evaluations. Phase 1. Toxicity Characterisation Procedures*, (EPA/600/3-88/034), US EPA, Duluth, MN, USA, 1988.
111. K. V. Thomas, R. E. Benstead, J. E. Thain and M. J. Waldock, *Mar. Pollut. Bull.*, 1999, **38**, 925.
112. K. V. Thomas, J. E. Thain and M. J. Waldock, *Environ. Toxicol. Chem.*, 1999, **18**, 401.
113. P. Matthiessen and R. J. Law, *Environ. Pollut.*, 2002, **120**, 739.
114. D. Schiedek, B. Sundelin, J. W. Readman and R. W. Macdonald, *Mar. Pollut. Bull.*, 2007, **54**, 1845.
115. BBC, *Eel Populations in London's River Thames Crash by 98%*, 2010, http://news.bbc.co.uk/1/hi/england/london/8473965.stm (last accessed 09/02/2011).

Harmful Algal Blooms

KEITH DAVIDSON,* PAUL TETT AND RICHARD GOWEN

ABSTRACT

Phytoplankton are free-floating plants found in marine and freshwaters that through their photosynthetic growth form the base of the aquatic food chain. A small subset of the phytoplankton may be harmful to human health or to human use of the ecosystem. The species that cause harm are now widely referred to as 'Harmful Algae' with the term 'Harmful Algal Bloom' (HAB) commonly being used to describe their occurrence and effects. In terms of human health, the most important consequence is the production, by some species, of biotoxins. Typically, biotoxin-producing phytoplankton species exist at relatively low densities (c. few hundred or thousand of cells per litre) with the toxins becoming concentrated in the flesh of organisms (particularly bivalve molluscs) that filter feed on phytoplankton. In most cases, there are no adverse effects to these primary consumers, but this concentrating mechanism creates a risk to health if the shellfish are consumed by humans. In this review, we provide an overview of the mechanisms through which marine phytoplankton may cause harm to humans in terms of heath, and the negative effects on the use of ecosystem services. Subsequently, we consider HAB issues in the area we are most familiar with: UK coastal waters. Finally, the methodologies used to safeguard human health from HAB-generated syndromes are discussed.

1 Phytoplankton

Phytoplankton is the collective name for the tiny organisms in lakes and seas. The word derives from the Greek *phyton* (plant) and *planktos* (wandering),

---

*Corresponding author

Issues in Environmental Science and Technology, 33
Marine Pollution and Human Health
Edited by R.E. Hester and R.M. Harrison
© Royal Society of Chemistry 2011
Published by the Royal Society of Chemistry, www.rsc.org

because these free-floating plants are transported throughout lakes and seas by currents and tides. A member of the phytoplankton is a 'phytoplankter'. Like plants on land, all phytoplankters contain the green pigment chlorophyll which enables them to photosynthesise organic matter from carbon dioxide, water and inorganic nutrients such as mineral salts containing nitrogen and phosphorus. These microscopic algae (micro-algae) represent less than 1% (*c.* 1 Pg C [1 Pg = 10^{15} g]) of the photosynthetic biomass of the Earth[1] but are responsible for about half of annual global net primary production. The phytoplankton therefore occupies a key position in determining global climate, and oceanic and atmospheric chemical composition.[2] Phytoplankton form the base of the marine and freshwater food chain, and the productivity of lakes, seas and oceans depends on them.

In temperate-latitude seas, the onset and duration of the phytoplankton production season is controlled by light availability.[3,4] This results in pronounced seasonality (see Figure 1) that is broadly determined by the cycle of solar radiation. During the production season, however, it is the supply of mineral nutrients that largely determines how much phytoplankton growth occurs. Nutrients accumulate during the winter period when there is insufficient light for growth and the supply of nutrients exceeds the demand by phytoplankton. When the light climate improves in late winter/early spring, these winter nutrients fuel spring growth. In general, these nutrients are not replenished and the low concentrations which prevail during the summer months constrain phytoplankton growth.

In other regions of the world, differences in the light climate and nutrient regimes result in different seasonal cycles of phytoplankton growth. For example, the South China Sea[5] and Mediterranean Sea[6] are naturally nutrient poor (oligotrophic) and, despite high levels of irradiance, the biomass of phytoplankton in near-surface waters is low. Other coastal regions (*e.g.* north western USA, north western Africa and Peru) are naturally nutrient enriched and productive as a consequence of upwelling, a process whereby deeper nutrient-rich water is entrained into surface illuminated waters and supports phytoplankton growth.[7]

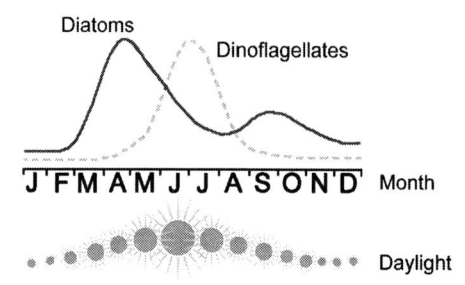

Figure 1 Schematic representation of the idealised annual cycle of phytoplankton in northern temperate coastal waters.

Human activities provide other sources of nutrients and the introduction of anthropogenic nutrients into coastal waters can stimulate phytoplankton production that, in turn, has undesirable effects on the ecosystem and its human use. This is the process of eutrophication.[8,9]

World-wide there are approximately 4000 species of phytoplankton.[10] These are commonly grouped into three functional categories or life-forms: diatoms, dinoflagellates and microflagellates. Diatoms have a silicon cell wall (or frustule) which is divided into two halves. The frustule often appears finely chiselled and may have spines or extensions. All diatoms are photosynthetic. Dinoflagellates have two flagella with which they can swim short distances. Some dinoflagellates are photosynthetic, but others are mixotrophic (able to photosynthesise and use organic matter as a source of energy) and others are heterotrophic (requiring organic matter as a source of energy). Microflagellates are a taxonomically and nutritionally diverse group of small (approximately ≤ 20 μm) organisms which have one or more flagella.

Populations of individual species are not fixed in time and space but are dynamic, particularly in coastal waters. Most phytoplankters typically reproduce by binary division and the normal pattern of growth involves an exponential increase in cells over a period of days. Under certain circumstances, therefore, the abundance of phytoplankton as a whole, or of one or more species, can increase rapidly. Such an occurrence is often referred to as a 'bloom'. Phytoplankton blooms are natural events. For example, the spring bloom that occurs in temperate waters each year[11,12] is an important part of the annual cycle of phytoplankton growth. A microscope image of a diatom-dominated phytoplankton bloom is presented in Figure 2.

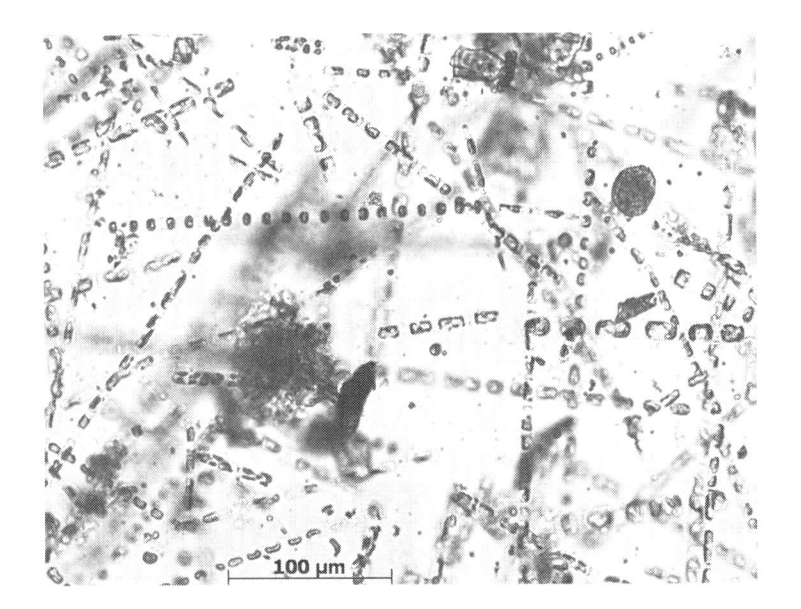

Figure 2 Diatom-dominated phytoplankton bloom.

1.1 Harmful Phytoplankton

On occasions, and under the right conditions, the abundance of cells can be so high that a bloom is visible to the human eye as a discolouration of the sea. In Japan, visible blooms associated with the death of fish and other marine animals were termed 'red-tides'.[13] The terms 'exceptional bloom'[14] 'nuisance' and 'noxious'[15] have also been used to describe both species and blooms which have a negative impact on ecosystem services. The species that cause harm are now widely referred to as 'harmful algae' and the term 'Harmful Algal Bloom' and the acronym 'HAB' is part of the scientific and colloquial language amongst scientists working in this field.

Approximately 300 species of phytoplankton have properties that make them harmful to humans or influence the human use of the ecosystem. Most typically this is from a human health perspective through the production of biotoxins. Other species may have a negative effect on ecosystem health or restrict the human use of the ecosystem by impacting on fisheries, aquaculture, recreational activities or nature conservation. As these latter impacts may have direct effects on human wellbeing, we also discuss the main causative organisms and their effects.

When considering harmful phytoplankton, a distinction needs to be made between harm which results from high and low biomass 'blooms'. The former is often the result of competition for oxygen, smothering or abrasion of fish gills. Generally, but not exclusively, problems associated with biotoxins are associated with low biomass blooms (approximately a few hundred or thousand of cells per litre).

The development of high biomass blooms requires a nutrient supply. This supply might be natural (as at the interface of mixed and stratified waters or coastal upwelling) or of human origin. The stimulation of the occurrence of HABs (where none have occurred before), or an increase in the frequency of occurrence, duration or spatial extent is considered to be one undesirable consequence of the human-driven eutrophication process.[16] However, whether or not anthropogenic nutrient enrichment of coastal waters influences the dynamics of HABs is a complex matter, with some of the key issues relating to this topic having recently been reviewed.[17]

Harmful phytoplankters are present within most phytoplankton classes. The majority of the biotoxin-producing species are members of the dinoflagellates, although one major toxic diatom genus (*Pseudo-nitzschia*) has also been identified. Some species of microflagellates also produce biotoxins (such as the fish-killing genus *Chatonella*). The majority of harmful phytoplankters are pelagic, although a number such as *Prorocentrum lima, Ostreopsis* spp. and *Gambierdiscus* spp. are semi benthic, living in the sediments or growing epiphytically on seaweed, and are only sporadically present in the water column. Harmful dinoflagellates and diatoms are most prevalent in marine environments, although harmful cyanobacteria (sometimes called 'blue green algae') can be important in particular environments such as the Baltic Sea. In this review we focus on marine micro-algae.

1.2 Mechanisms of Harm to Human Health

Phytoplankton can cause harm to human health by a number of routes. As noted above, the most prevalent is through exposure to biotoxins. A range of toxins exist with differing modes of action. Exposure mechanisms also vary for different toxins. Three different mechanisms may cause these biotoxins to influence human health: direct ingestion/contact, aerosolised transport, or concentrating and vectoring by a marine organism that is then eaten (typically a form of shellfish).

Most of the known algal toxins are neurotoxins, although some can cause liver damage and cancer. The majority of human diseases associated with exposure to biotoxins appear to be acute phenomena, although some (*e.g.* ciguatera fish poisoning) can cause prolonged sub/chronic disease.[18]

1.3 The Scale of the Problem

Worldwide, fish and shellfish consumption is expanding. Globally, wild fish stocks are in decline and aquaculture is rapidly growing in importance.[19,20] On a regional scale the contribution of shellfish aquaculture to marine food consumption is variable. Shellfish consumption dominates in parts of Asia. European *per capita* consumption is considerably lower and is itself quite regional in pattern. In the UK for example, shellfish production is relatively small. In 2006, farmed shellfish production totalled 15 449 tonnes and was dominated by cultivated mussels (14 553 tonnes) and oysters (880 tonnes).[21] However, in Scotland where 80% of rope-grown (rather than dredged) production occurs, shellfish is regarded as a premium product with significant local and export markets that would easily be damaged by major human health incidents related to biotoxin-contaminated seafood. Moreover, the socio-economic importance of shellfish harvesting is magnified by the fact that in the UK, and many other parts of Europe, production is often located in rural areas and contributes disproportionately to the local economy.

Aquaculture production will continue to support the increasing human demand for seafood to feed both mass and quality markets in different regions of the world. As coastal aquaculture, and shellfish production in particular, is vulnerable to contamination by biotoxins, it is likely that the potential human health effects of these compounds will also continue to increase.

2 Human Health Syndromes

2.1 Shellfish Poisoning

Within the phytoplankton and the water column as a whole, algal biotoxins are typically at such low concentrations that they pose no threat to humans through direct contact or ingestion (although see the discussion of aerosols below). Through their filter feeding on phytoplankton, shellfish, in particular bivalve molluscs, are exposed to biotoxins. Following ingestion, biotoxins can accumulate within the shellfish flesh. In most cases there are no adverse effects

Table 1 Regulatory methods of shellfish toxicity determination in UK waters.

Toxin group	Method	Regulatory limit
Amnesic shellfish poisoning	HPLC	20 μg domoic acid per g of shellfish flesh
Paralytic shellfish poisoning	HPLC pre-screen Mouse bioassay (MBA) quantification for positive samples	80 μg saxitoxin equivalence per 100 g of shellfish flesh
Lipophilic toxins	MBA*	Presence by MBA

*Currently DSP toxins are detected using a mouse bioassay (MBA). However, it is expected that the MBA for lipophilic toxins will be phased out within the next year and be replaced by an analytical LC-MS methodology with quantifiable limits per weight of shellfish flesh.

to the shellfish themselves, but this concentrating mechanism creates a risk to human health if the shellfish are consumed. Some toxins can remain within the tissue of shellfish for considerable lengths of time after the abundance of the causative phytoplankton has decreased. Furthermore, as phytoplankton-derived shellfish toxins are generally unaffected by cooking, human health can only be safeguarded by preventing the ingestion of contaminated shellfish. Toxicity of shellfish (in UK waters) is determined either by biological assay (using mice) or analytical techniques based on high-performance liquid chromatography (HPLC), as shown in Table 1.

2.2 Causative Organisms and Toxins

2.2.1 Paralytic Shellfish Poisoning (PSP)

The toxins that cause paralytic shellfish poisoning (PSP) are derivatives of the alkaloid saxitoxin. At least 21 derivatives of this compound exist[22] with various concentrations and combinations having been associated with PSP.[23] Paralytic shellfish toxins (PSTs) are highly lethal, with an LD_{50} (median lethal dose) in mice by intra-peritoneal injection of 10 μg kg^{-1}, 1000 times more toxic than sodium cyanide[24] that has an LD_{50} of 10 mg kg^{-1}. The mechanism of saxitoxin toxicity is to act as a sodium-channel blocker, inhibiting the transmission of action potentials in nerve axons and skeletal muscle fibres.

In addition to shellfish gastropods, crustacean and fish are also possible vectors.[25] Paralytic shellfish, poisoning may result in severe gastrointestinal symptoms. Neurological symptoms include facial tingling and numbness. This can develop into extensive muscular paralysis and, in severe cases, lead to death.[26,27] The IOC-UNESCO (Intergovernmental Oceanographic Commission-United Nations Educational, Scientific and Cultural Organisation) Harmful Algal Bloom Programme indicates that there are about 2000 cases of PSP annually with a 15% mortality.

The armoured chain-forming dinoflagellate, *Pyrodinium bahamense* var. *compressum*, was responsible for the greatest number (41%) of global paralytic shellfish poisoning events between 1989 and 1999 (ref. 28). This organism, first described in 1906, is of importance in several countries in the tropical Pacific, with the harmful implications of the organism being first recognised in Papua

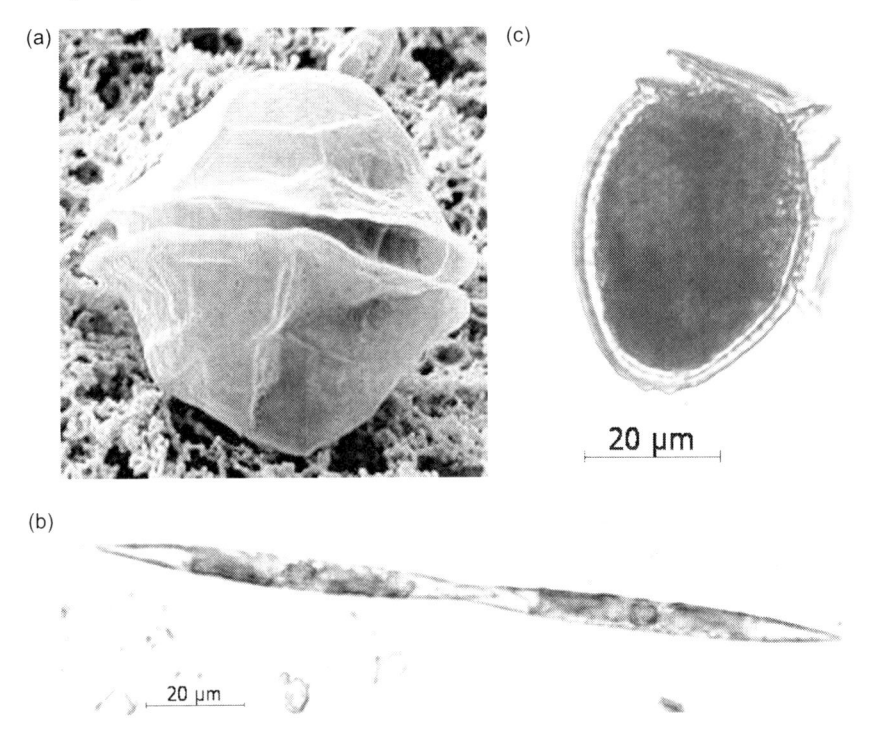

Figure 3 (a) *Alexandrium tamarense*. (b) *Pseudo-nitzschia seriata* group. (c) *Dinophysis acuminata*.

New Guinea in 1972. Blooms are often related to monsoon periods with wind-driven upwelling being implicated in their formation,[29] potentially through re-suspension of cysts into the water column where excystment and vegetative cell division results in population growth.

Other important producers of PSTs include dinoflagellates of the genus *Alexandrium;* see Figure 3(a). The genus includes over 25 morphologically described species with some having both toxic and non-toxic strains.[30] Those species that produce toxins often have very different toxin levels and profiles. The majority of toxicity events are caused by the species *A. tamarense*, *A. catanella* and *A. fundyense*[31] which make up the *A. tamarense* species complex, the biogeography of which has been reviewed by Lilly *et al*.[31] The first PSP event was reported in 1927 near San Francisco, USA, and was caused by *Alexandrium catenella*. This event resulted in 102 people being ill and six deaths.[32] *Gymnodinium catenatum* is also an important PST-producing organism, especially on the Iberian Peninsula,[33] and may have been present in Europe since at least the 1600s (ref. 34).

2.2.2 Amnesic Shellfish Poisoning (ASP)

Marine diatoms of the genus *Pseudo-nitzschia* Peragallo, as shown in Figure 3(b), produce the neurotoxin domoic acid (DA)[35] that can result in amnesic shellfish poisoning (ASP). The genus contains at least 19 species with approximately half of these having been found capable of producing DA. *Pseudo-nitzschia* is near unique amongst the approximately 2000 described diatom species in its ability to produce a biotoxin.[36]

Domoic acid is a low-molecular-weight, water-soluble tricarboxylic acid. Amnesic shellfish poisoning is a result of DA mimicking the neurotransmitter glutamic acid. Domoic acid binds irreversibly to glutamate receptor sites in the brain, causing massive depolarisation of neurons, increasing cellular Ca^{2+} which, in turn, induces neuron swelling and death.[22] As the affected nerve cells are located in the hippocampus and function in memory retention, memory loss may occur.[35] The symptoms of ASP are nausea, vomiting, abdominal cramps, diarrhoea, memory loss, decreased consciousness, seizures, confusion and disorientation.[24]

The first documented ASP event occurred relatively recently in Prince Edward Island (Nova Scotia, Canada) during November/December 1987. Domoic acid contamination of blue mussels resulted in at least three deaths and in excess of 100 people suffering illness. The causative organism of this ASP event was *P. multiseries* (Hasle) Hasle.[37] Subsequently, blooms of these phytoplankters have resulted in a range of, often larger scale, shellfish toxicity events. Examples include: *Pseudo-nitzschia australis* that caused ASP symptoms, including death, in sea mammals and birds on the California coast in Monterey Bay,[38] *Pseudo-nitzschia calliantha* (previously referred to as *P. pseudodelicatissima*) closed shellfish harvesting in the Bay of Fundy in 1988 (ref. 39), and *Pseudo-nitzschia seriata* that was associated with elevated DA levels in molluscs in the Gulf of St. Lawrence, Canada, between 1998 and 2000 (ref. 40), and in 2002 (ref. 41).

As DA is an amino acid, it contains nitrogen; hence a possible explanation for the production of DA is as a secondary metabolite repository for excess nitrogen when other nutrients are in lower relative supply. Consistent with this view, Pan *et al.*[42] found that silicate limitation increased DA production by *Pseudo-nitzschia multiseries*. Domoic acid was produced when population growth was declining and was at a maximum when cells were silicate depleted, consistent with field observations made during the first DA incident in Prince Edward Island when domoic acid production peaked 10 days after the bloom maximum, and when silicate in the water was depleted. Similarly, Fehling *et al.*[43] reported that *Pseudo-nitzschia seriata* produced DA in both phosphorus- and silicate-limited conditions, but that more domoic acid was generated in the latter condition. However, an alternative explanation for the DA production was presented by Rue and Bruland[44] who demonstrated DA to be capable of mediating the acquisition (iron) or detoxification (copper) of trace metals in seawater. A range of further questions also remain unanswered with respect to *Pseudo-nitzschia* and its toxicity. Perhaps most prominent amongst these is the perceived widespread increase in ASP events: it remains unclear why ASP was not observed anywhere worldwide prior to 1989 but that there have been frequent observations since.

2.2.3 Diarrhetic Shellfish Poisoning (DSP)

This syndrome is caused by dinoflagellates from the genera *Dinophysis*, as shown in Figure 3(c) and by *Prorocentrum lima*. Diarrhetic shellfish poisoning was first linked to the presence of *Dinophysis acuminata* in Dutch coastal waters in 1961 (ref. 45) and *D. fortii* in Japan.[46]

The DSP toxin group is made up of four main lipophilic shellfish toxin compounds (LSTs): okadaic acid (OA) and its derivatives, the dinophysistoxins (DTXs) – DTX-1, DTX-2 and DTX-3. However, new OA-class toxins continue to be discovered. The chemical and physical properties of the DSP toxin group were reviewed by Quilliam and Wright.[47]

In humans, DSP toxins bind to protein phosphatase receptors, building up phosphorylated proteins.[22] While generally less serious than PSP or ASP, DSP is still characterised by severe gastrointestinal symptoms with the rapid (within several hours) onset of diarrhoea, vomiting and abdominal cramps. Diarrhetic shellfish poisoning is not fatal, and those affected usually recover within a few days. There are many reported incidents worldwide each year, but as symptoms are not always reported or are misdiagnosed as a stomach disorder, the true magnitude of DSP-related illnesses remains unquantified. While the effects of DSP are generally short-lived, chronic exposure is suspected to promote digestive system tumour formation.[24]

Dinophysis is a large genus with approximately 200 described species. Diarrhetic shellfish toxin production has been confirmed in eight of these, with two others being suspected.[22] *Dinophysis* may be aggregated in coastal regions due to wind-driven exchange[48,49] but rarely reaches concentrations sufficient to discolour seawater.

Diarrhetic shellfish poisoning outbreaks are common in Europe, with at least 10 countries having experienced DSP events.[22] The most widespread events were in 1981 and 1998 in Spain and France, respectively, with the number of recorded illnesses reaching into the thousands.[50,51] Diarrhetic shellfish poisoning has been of particular concern in Ireland, with harvesting closures of up to 10 months in duration occurring on some occasions.[52,53]

Study of the factors governing the toxicity of *Dinophysis* is less advanced than for *Alexandrium* or *Pseudo-nitzschia*. This is because, until recently, it was not possible to maintain the organism in laboratory culture for extended periods of time to conduct the necessary experiments. This problem was recently solved by Park *et al.*[54] who demonstrated the mixotrophic nature of *Dinophysis* and maintained it in culture using a predator/prey system involving a ciliate prey (*Myrionecta rubra*).

Prorocentrum lima is also associated with DSP, with Pan *et al.*[55] reporting the production of OA, OA diol ester, DTX1 and DTX4 by this species. While not as frequently associated with DSP as *Dinophysis*, the semi-benthic nature of *P. lima* may lead to its under-sampling by monitoring programmes, perhaps leading to underestimates of its importance.

2.2.4 Other Lipophilic Shellfish Toxins (LSTs)

Pectenotoxins (PTXs): In addition to their production of OA and DTXs, *Dinophysis* species are linked to the production of PTXs, a group of polyether-lactone toxins. Their presence in shellfish was discovered due to their acute toxicity in the mouse bioassay after intra-peritoneal injections of lipophilic shellfish extracts. It is important to note that there are no reports on adverse effects in humans associated with the PTX group of toxins, but since PTXs always co-occur with OA toxins it has been difficult to assess whether the PTX group of toxins alone contribute to incidences of human illness.

Yessotoxin (YTX): The dinoflagellates, *Lingulodinium polyedrum, Protoceratium reticulatum* and *Gonyaulax grindleyi*, have been implicated in the production of YTX, a marine polyether toxin that was first isolated in 1986 in Japan. It has since been found at various sites in Europe, North and South America, and New Zealand. In the UK, analysis of shellfish from Scotland using liquid chromatography-mass spectrometry (LC-MS) has demonstrated the presence of multiple LSTs at low concentrations.[56]

Yessotoxins accumulate in shellfish and are toxic to mice by intra-peritoneal injection, producing symptoms similar to those of PSP toxins. Yessotoxins have been associated with diarrhetic shellfish poisoning (DSP) because they are often simultaneously extracted with DSP toxins. However, recent evidence suggests that YTXs should be excluded from the DSP toxins group because, unlike okadaic acid (OA) and dinophysistoxin-1, YTXs do not cause either diarrhoea or inhibition of protein phosphatases. Yessotoxin are potent cytotoxins and this has led European Authorities to establish a maximum permitted level in shellfish of 1 mg YTX equivalents kg^{-1} (ref. 57).

Azaspiracid (AZA): In November 1995, at least eight people in the Netherlands became ill after eating mussels from Ireland. Although symptoms resembled those of DSP, concentrations of the major DSP toxins were low[58] and no known DSP-producing species were observed in the water. In addition, the neurotoxic symptoms exhibited in mouse bioassays were quite different from typical DSP toxicity, with a slowly progressing paralysis being observed.[59] This event led to the 'discovery' of AZA toxins.

Azaspiracids differ structurally from any of the previously known marine toxins and belong to a novel group of polyethers.[60] They have unique spiro ring assemblies, a cyclic amine instead of a cyclic imine group, and the absence of a carbocyclic or lactone ring.[61] Azaspiracid toxicity has now been demonstrated in a number of other EU countries with approximately 20 different analogues of AZA having been identified. Based on occurrence and toxicity, the European Food Safety Authority considers AZA1, AZA2 and AZA3 to be most important. Azaspiracid toxicity was initially thought to be linked to the heterotrophic dinoflagellate, *Protoperidinium crassipes*, but this proved to be incorrect. However, recently a small dinoflagellate, first isolated from the East coast of Scotland, *Azadinium spinosum*, has been identified as an AZA producer.[62]

2.2.5 Neurotoxic Shellfish Poisoning (NSP)

This syndrome is generated by the brevetoxin group of toxins. Brevetoxins are of particular importance in Florida and the Gulf of Mexico where they are primarily produced by the naked dinoflagellate, *Karenia brevis*.

Brevetoxins activate voltage-sensitive sodium channels, causing sodium influx and nerve membrane depolarisation. Neurotoxic shellfish poisoning (NSP) exhibits both gastrointestinal and neurological symptoms with nausea and vomiting, paresthesias (tingling, pricking, or numbness) of the mouth, lips and tongue, distal paresthesias, ataxia (loss of ability to coordinate muscular movement), slurred speech and dizziness.[63] Partial paralysis and respiratory distress may also occur. Although no NSP-induced fatalities have been recorded,[64] symptoms may be severe enough to require hospitalisation.

Blooms of *K. brevis* can reach a size and cell density that discolour the water and, in addition to human health problems, has resulted in massive fish kills in the Gulf of Mexico, Florida and North Carolina coastal regions.[65]

2.3 Respiratory Illness

In addition to the shellfish vectoring described in the previous section, it has now been demonstrated that *Karenia brevis*-produced toxin can be transported as an aerosol. Reports of respiratory problems in asthmatics have been linked to red tides of the phytoplankter.

Respiratory illness has also been linked to the benthic dinoflagellate, *Ostreopsis ovata*. In tropical seas, *O. ovata* is often associated with the genera *Gambierdiscus*, *Coolia* and *Prorocentrum*, causing ciguatera fish poisoning[66] (see section 2.4). However, the health implications of *O. ovata* have relatively recently become evident in Europe in the Mediterranean and Adriatic seas. *Ostreopsis ovata* produces palytoxin which targets the sodium-potassium pump protein by binding to the molecule such that the molecule is locked in a position where it allows passive transport of both the sodium and potassium ions, thereby destroying the ion gradient that is essential for most cells.

One of the first European reports was from Genova, Italy, in 2005, where people who spent time on or near beaches, experienced symptoms such as rhinorrhoea (runny nose), cough, fever, bronchoconstriction with breathing difficulties, wheezing, and conjunctivitis.[67] In general, the symptoms ceased after a few hours, but 20 people required hospital treatment. The event was associated with high densities of *O. ovata* and laboratory tests demonstrated the presence of putative palytoxin.[68]

2.4 Fish Vectored Illness

Related to brevetoxins are the ciguatoxins (CTX). These toxins arise from biotransformation in fish of gambier toxins produced principally by the benthic dinoflagellates, *Gambierdiscus toxicus*, although a range of other organisms, including a species of *Prorocentrum*, *Gymnodinium sanguineum* and *Gonyaulax polyedra*, have also been implicated.[69] *Gymnodinium toxicus* live in association with seaweeds, sediments and coral rubble. On ingestion by herbivorous fish, the toxins are concentrated and then passed up the food chain to larger carnivorous fish. During this process of trophic transfer there is an oxidative transformation of the less oxidised gambier toxins to more oxidised and more toxic CTX that generates the syndrome 'Ciguatera Fish Poisoning' (CFP) in humans.[70]

Ciguatoxin activates voltage-gated sodium channels in cell membranes, increasing sodium ion permeability, resulting in depolarisation of the nerve cell, producing a range of gastrointestinal, neurologic and/or cardiovascular symptoms which last days to weeks, or even months.[18] While CTX is a potent toxin that has caused human mortalities, it is fortunate that it rarely accumulates in fish to levels that are lethal to humans.[71] However, it remains the most common type of marine food poisoning worldwide with an estimated 10 to 50 thousand people suffering from the disease annually in areas of the Pacific Ocean, Caribbean Sea and Western Indian Ocean.[72]

2.5 Cyanobacteria

There are approximately 2000 species of cyanobacteria, of which 40 have been identified as toxigenic,[73] with toxicity to hepato-, neuro-, gastro-intestinal and dermatic systems with embryo-lethal, teratogenic, gonadotoxic, mutagenic and tumour-promoting activities.[74] Cyanobacteria and their toxins are of perhaps greatest concern in freshwater environments and low salinity marine waters such as the Baltic Sea.[75]

Important among the marine cyanobacteria is the benthic species *Lyngbya majuscula* that can be found in tropical regions growing in fine strands attached to seaweed and rocks. Mats can rise to the surface to form large floating aggregations that may reach land. *Lyngbya majuscula* contain a number of toxic compounds of which debromoaplysiatoxin (DAT), aplysiatoxin (AT) and the lyngbyatoxins A, B and C are the most important. It is responsible for cyanobacterial dermatitis, commonly referred to as 'swimmers' itch' or 'seaweed dermatitis'. This is a severe contact dermatitis that may occur after swimming in water. The symptoms are itching and burning within a few minutes to a few hours after swimming. Visible dermatitis, blisters and deep desquamation may follow, with eye and respiratory irritation also possible. Toxins from the genera *Schizothrix* and *Oscillatoria* have also been linked to dermatitis and tumor formation.

A further important toxic species is *Nodularia spumigena*, which is of particular concern for human and animal health in the Baltic Sea.[76] This species produces nodularin, a monocyclic penta-peptide (closely related to the freshwater microcystins) that acts as a hepatotoxin, by inhibiting protein phosphatase activity. Low-level exposure to these toxins may promote the development of cancer in the liver and other chronic disorders of the gastrointestinal tract. Allergic or irritative dermal reactions have also been reported following recreational exposure.

2.6 The Role of Harmful Phytoplankton in Influencing Human Wellbeing

A number of phytoplankton species have other negative effects on the goods and services that marine ecosystems provide for humans. Their effects relate

Harmful Algal Blooms 107

principally to farmed fish but may extend to their influencing the recreational use of coastal environments.

2.6.1 Microflagellates

Blooms of microflagellates have caused mortalities of farmed fish throughout the world. In Northern Europe, Kaartvedt et al.[77] reported a bloom of the toxin-producing *Prymnesium parvum* in a Norwegian fjord system which caused mortalities of farmed fish with an economic loss of US$5 million. In Japan, blooms of several microflagellate species have been associated with fish kills. *Chattonella antiqua* blooms have frequently caused fish kills and, in 1972, a bloom was associated with the mortality of c. 14.2 million Yellowtail with an economic loss of US$70 million.[78] Blooms of *Heterosigma akashiwo* caused serious damage to the aquaculture industry in the Seto Inland Sea during 1975 and 1981 (ref. 79) and have caused mortalities of farmed fish in coastal waters of the Canadian Pacific[80] and in New Zealand.[81]

A group of phenomena called 'foaming of the sea' or 'mucilage events' in coastal waters are often associated with organic matter produced by certain phytoplankters, particularly of the genus *Phaeocystis*. Species of *Phaeocystis* have a colonial life stage during which large numbers of cells are embedded in a mucilaginous matrix up to 2 cm in size.[82] During *Phaeocystis* blooms, agitation of the water by wind and wave action can break up the colonies and the mucilage creates large quantities of foam. Typically these events appear as masses of wind-blown foam, initially accumulating in persistent white drifts on beaches and then decaying to "impressive layers of brownish and stinking foams".[83]

Like other prymnesiophytes, *Phaeocystis* plays an important part in the global sulfur cycle.[84] It produces and releases 3-dimethylsulfoniopropionate (DMSP), the precursor of dimethyl sulfide (DMS), and it is, presumably, the latter compound that contributes to the 'rotten-cabbage' smell of decaying blooms and foam. Savage[85] reported that North Sea fishermen held that fish avoided *Phaeocystis* blooms and associated 'stinking water'. More recently, blooms of *Phaeocystis* have been reported to clog fishing nets and interfere with commercial fishing.[86]

Species of *Phaeocystis* have also been associated with mortalities of farmed fish. In September 1997, for example, a bloom of *P. globosa* in Quanzhan Bay, Fujian province (China), extended over an area of 3000 km^2 and caused major losses of farmed fish estimated as US$7.5 million.[87] Experiments with fish larvae[88] suggested that water that had contained the algae was poisonous to young cod, although the harmful substance might have had more of an anaesthetic effect rather than the haemolytic activity found from material obtained from other prymnesiophytes.[89]

There are no reports of *Phaeocystis* or *Phaeocystis*-made substances being directly harmful to humans, and thus the main route to impact on human well-being is likely to be the economic effect of beach foam in deterring tourism.

2.6.2 Other Dinoflagellates

In this chapter it is not possible to review all harmful dinoflagellates. However, some important examples are mentioned below.

Noctiluca scintillans is an important red-tide organism world-wide. It occurs in two forms: red *Noctiluca* is heterotrophic and fills the role of one of the microzooplankton grazers in the food web; in contrast, green *Noctiluca* contains a photosynthetic symbiont, *Pedinomonas noctilucae*, but it also feeds on other plankton when the food supply is abundant.

Species of the genus *Karlodinium* are of concern due to their ichthyotoxic properties. For example, blooms of *Karlodinium veneficum* (synonyms *Karlodinium micrum* and *Gyrodinium galatheanum*) have been evident since 1950 in a range of coastal waters including Southwest Africa, Europe, United States and Western Australia, often coinciding with fish kills.[90] Laboratory study has demonstrated the production of lipophilic karlotoxins that interact with certain sterols to increase ionic permeability of cell membranes.[25] Resultant osmotic cell lysis can cause damage of the gill epithelia resulting in hypoxia or death.

Cochlodinium polykrikoides blooms have been reported from several coastal regions of Japan, including a large bloom in 2000 which caused losses of *c.* US$36.4 million.[91] This species has been associated with fish kills and mortalities of coral reef organisms in Pacific coastal waters of Costa Rica[92] and mortalities of fish in South Korea.[93]

Blooms of the naked dinoflagellate *Karenia mikimotoi* (previously named *Gyrodinium aureolum*) have caused mass mortality of farmed fish in Norway,[94] Ireland,[95] Scotland,[96] China (Hong Kong)[87] and South Korea.[93] Many of the early reports of mortalities of benthic animals and farmed fish were attributed to de-oxygenation; however, toxin-like damage to fish gills has been reported,[97] and cytotoxins have been identified.[98]

2.6.3 Diatoms
Species of diatom belonging to the genera *Chaetoceros* and *Skeletonema* are a common component of the marine flora in coastal waters throughout the world, usually without harmful effects, but occasional blooms have resulted in losses of farmed fish. For example, *Chaetoceros* sp. have been associated with mortalities of farmed fish on the Pacific coast of Canada and in Scotland.[99] The cause of death is presumed to be gill damage (from the siliceous spines of the cell wall), leading to asphyxiation.

3 Harmful Algal Blooms in UK Coastal Waters

Below, as a case study, we present a review of harmful algal blooms (HABs) in UK coastal waters and their influence on human health and wellbeing.

3.1 Shellfish Poisoning

In UK waters we are fortunate that the number of toxin-producing phytoplankton is limited, with eight species or genera being monitored on a routine basis to safeguard human health (see Table 2). Of these, species of dinoflagellates belonging to the genera *Alexandrium* and *Dinophysis* and diatoms

Table 2 Species or genera of potentially biotoxin-producing phytoplankton monitored in UK waters to ensure shellfish safety.

Organism	Toxin: syndrome
Alexandrium spp.	Saxitoxin and derivatives: PSP
Pseudo-nitzschia spp.	Domoic acid: ASP
Dinophysis spp.	Okadaic acid and dinophysistoxins: DSP
Prorocentrum lima	Okadaic acid and dinophysistoxins: DSP
Prorocentrum minimum	Venerupin: VSP
Lingulodinium polyedrum	Yessotoxin: YTX
Protoceratium reticulatum	Yessotoxin: YTX
Protoperidinium crassipes	Azaspiracid: AZP

belonging to the genus *Pseudo-nitzschia* are the most important. While other lipophilic shellfish toxins (LSTs), including YTXs, PTXs and AZAs, have been detected in shellfish from UK waters, these have been relatively infrequent and at low levels. Whether this is related to less rigorous monitoring for these toxins is unclear.

It is evident that some of the species that are now regarded as harmful have been a natural part of the phytoplankton in UK coastal waters for at least the last 100 years, and hence it is important to note that HABs are not a new phenomenon. For example, Cleve[100] reported *Dinophysis acuta* in the northern North Sea off Scotland and in the Irish Sea. Herdman and Riddell[101,102] recorded the presence of *Dinophysis* sp. in the Scottish west coast sea Lochs Hourn (in July 1908 and 1909) and Torridon (in July 1911), and in the Firth of Lorne in 1909. In addition, Lebour[103] recorded the presence of *Dinophysis acuminata* in the English Channel (off Plymouth) during a study in 1915 and 1916.

With respect to other important genera, *Alexandrium tamarense* (or *Gonyaulax tamarensis* as it was first named), the original description was based on cells collected from the Tamar estuary.[104] For *Pseudo-nitzschia*, the presence of *Nitzschia seriata* (now called *Pseudo-nitzschia seriata*) was recorded in Scottish waters in Loch Hourn in July 1909; in the Firth of Lorne in 1909 and 1910; and both *N. seriata* and *Nitzschia delicatissima* (the latter up to 335×10^6 cells l^{-1} based on a net sample) in Loch Torridon in July 1911 (ref. 101,102).

3.1.1 Paralytic Shellfish Poisoning (PSP)
In UK waters the organism of most concern is *Alexandrium tamarense* (Group I) which has historically been viewed as a potent PSP-producer in Scottish waters, with particular hotspots including the Orkney and Shetland Islands. It is important to recognise that *A. tamarense* often constitutes a very minor component of the phytoplankton assemblage but, through its very high per-cell toxicity, remains capable of generating shellfish toxicity even at very low cell densities of only a few hundred cells per litre. Non-toxic forms of *A. tamarense* (group III) have also been identified in UK waters, with co-occurrence of groups I and III now having been demonstrated.[105,106]

The less toxic *Alexandrium minutum* has been identified as a PSP-producer in England, Ireland and occasionally in Scotland. Its role as a PSP-causing organism is variable and cell densities rarely reach sufficient levels for PSP to cause major problems in the UK. However, significant localised problems occur in Cork Harbour in Southern Ireland.[107] *Alexandrium ostenfeldii* from both Scottish and English waters have been observed to produce trace amounts of PSP, as well as spirolides.

In the UK, PSTs have occurred in both western and eastern Scotland and the Northern Isles,[108,109] Belfast Lough in Northern Ireland,[110] and the North East of England,[111] but are relatively rare elsewhere. It is difficult to accurately assess the historical occurrence of PSP events in the UK as detailed records are not available before the 1990s. However, the earliest suspected recorded event is from the consumption of mussels in the Firth of Forth in 1827 (ref. 112,113). Based on available accounts, Ayres[114] suggested that between 1827 and 1968 there were ten PSP incidents, with approximately 14 fatalities. The first well-documented case of PSP intoxication in UK waters was in 1968 in the North East of England when 78 people showed clinical symptoms of PSP toxicity after consuming mussels[115] contaminated by *A. tamarense*.[116] At this time there were coincident reports of sea bird and sand eel mortalities.[117] It was also estimated that 80% of the breeding population of shags (*Phalacrocorax aristotelis*) in Northumberland died during this event.[118]

Between 1968 and 1990 there were no recorded clinical cases of PSP although threshold levels were breached 17 times,[108] with widespread toxicity being recorded in 1990 in the north east of England and in Belfast Lough, Northern Ireland, with toxicity reaching 3647 µg saxitoxin per 100 g shellfish tissue in mussels from Trow Rocks on the north east coast of England.[119]

3.1.2 Amnesic Shellfish Poisoning (ASP)

In UK waters, thirteen *Pseudo-nitzschia* species are known to be present.[120,121] Of these, only three are confirmed toxin producers, *P. australis* and *P. seriata* in Scottish waters, along with *P. multiseries* in English waters.

Prior to 1998, there are no records of ASP in UK waters. However, in 1999, a 49 000 km^2 area in western Scottish waters was closed to shellfish harvesting as a result of high levels of DA in king scallops. Scallop harvesting was first prohibited in June of that year, and the ban remained in force until April 2000. This is the largest fishery closure recorded worldwide.[122] Offshore and inshore waters were affected and this event demonstrates the potential significance of HAB-generated toxicity incidents in the UK. Frequent fishery closures occurred in subsequent years in the region due to elevated toxicity in scallops. The finding that the bulk of the toxin (typically >99% of DA) is located in the scallop hepatopancreas has led to a system of shucking and end-product testing of the edible gonad and muscle. This safeguards human health while allowing the offshore scallop fishing industry to be sustained.

While the end-product testing of scallops has markedly reduced the human ASP heath risk in UK waters, a risk still exists from the potential human

Harmful Algal Blooms

toxification from mussels in particular, as DA accumulates (and depurates) rapidly in mussel flesh.[123]

3.1.3 Diarrhetic Shellfish Poisoning (DSP)

The most commonly observed *Dinophysis* species in UK coastal waters are *D. acuminata* and *D. acuta*, but with considerable inter-annual variation in their absolute and relative abundances. For example, *D. acuta* dominated in Scottish waters in 2001 but numbers have subsequently declined as *D. acuminata* has become abundant (E. Bresnan, personal communication).

A number of incidents linked to DSP toxins have been recorded in UK waters. In June 2006, 171 people became ill after consuming mussels from a Scottish site and in 2007, an extended period of DSP toxicity (June to August) was recorded in shellfish from the Shetland Islands.[124] A number of human health problems have been associated with DSP in England and Wales but there are no reports in the scientific literature linking *Dinophysis* spp. to these incidents. While the number of reported events seems very few, it is likely this is due to huge under-reporting.

A second potential causative organism of DSP in UK waters is *Prorocentrum lima*. As noted above, its epiphytic behaviour means that this species may be under-represented in sampling programmes that better sample those species that are pelagic in nature. Scientific studies are relatively few, although the production of OA and DTX-1 from *P. lima* strains isolated from the Fleet Lagoon in Dorset, an area that had previously experienced DSP in shellfish, was demonstrated without any observable *Dinophysis* in the water column.[125]

3.1.4 Azaspiracid Poisoning (AZP)

Azaspiracids are currently monitored in shellfish using the mouse bioassay for DSP; the bioassay does not discriminate between different forms of lipophilic toxins, and hence the prevalence of azaspiracids in UK waters is not known. However, in a recent study of Loch Fyne in the west of Scotland[126] that characterised toxin profiles by LC-MS, azaspiracids were detected at low levels. Moreover, while the AZA producer *Azadinium spinosum* is known to exist in UK waters,[62] little is known about its spatial or temporal distribution, partly because it is not easily identified by light microscopy and molecular probe techniques that may allow its routine identification are not yet routinely available.

3.2 Other Harmful Phytoplankton in UK Waters

A number of other harmful phytoplankters occur in UK waters, with the potential to cause harm to the human use of ecosystem goods and services.

3.2.1 Karenia mikimotoi

Dense blooms of *Karenia mikimotoi* have had important environmental and economic consequences, with kills of farmed fish and benthic organisms being reported though water column deoxygenation. However, mortalities of farmed fish and wild organisms may be due to toxins produced by this phytoplankter,[98] although there are no reported human health implications.

Karenia mikimotoi at harmful densities was reported in UK waters, including the eastern Irish Sea, in 1971 and along the south western coast of England in 1975. Data from Scottish HAB monitoring programmes suggests that *Karenia mikimotoi* is found as a regular component of the phytoplankton, normally reaching only a few thousand cells per litre. However, harmful blooms have been recorded in 1980, 1999, 2003 and 2006 (ref. 96). The 2006 bloom was the most extensive to date and covered an area that extended from the island of Mull on the west coast, to the Shetland Isles and the north east coast. Abundance over the region varied but reached 3.7×10^6 cells l^{-1} in Scapa Flow in the Orkney Isles in mid August. Although no major fish kills were reported, there were reports of mortalities of benthic organisms including lugworm (*Arenicola marina*), blue mussel (*Mytilus edulis*), common starfish (*Asterias rubens*) and king scallop (*Pecten maximus*). There were also reports from the public of mortalities of crab and lobster, and fish including sea scorpion (*Myoxocephalus scorpius*) and conger eel (*Conger conger*).[96]

3.2.2 Other Dinoflagellates
A bloom of the dinoflagellate *Heterocapsa triqueta* (1.0×10^6 cells l^{-1}) in the Shetland Islands in May 2001 caused substantial losses to fish farms, as did a bloom (c. 9×10^6 cells l^{-1}) of an unidentified species of *Gymnodinium* in the Orkney and Shetland Islands in August of that year.[127]

3.2.3 Phaeocystis
As in the case of some of the toxin-producing species, historical accounts of *Phaeocystis* blooms in UK waters show that these events are not a new occurrence. Orton[128] described the aesthetic effect of *Phaeocystis* on the oyster beds of the Thames as 'baccy [tobacco] juice' and Savage[85] wrote that in April 1926 the main feature of the plankton was the presence of large quantities of *Phaeocystis pouchetii* that clogged the plankton nets.

Blooms of *Phaeocystis* spp. were recorded in the eastern Irish Sea in 1957 and 1958 (ref. 129). In 1992, a bloom (12 000 colonies l^{-1}) in the same area was linked to the deaths of fish and crustaceans, possibly as a result of oxygen depletion.[130] Blooms of *Phaeocystis* spp. in the English Channel are described as being annual events[131] and can often be dense and persistent.[132] In 2005, a *Phaeocystis* spp. bloom in the Shetland Islands that extended down the east coast of Scotland was also described as having an effect on farmed fish.[133] The occurrence of *Phaeocystis* spp. are recorded as part of a number of monitoring programmes in the UK[9] but, in general, recorded coastal blooms of *Phaeocystis* are infrequent.

3.2.4 Microflagellates
Records of harmful microflagellate blooms in UK waters are rare in the literature. One of the best documented are the blooms of an unidentified species

known as 'Flagellate X' (possibly *Heterosigma akashiwo*) in Loch Striven in 1979 (ref. 134) and again in 1982 in Loch Fyne.[135] These events were associated with mortalities of farmed fish.

3.2.5 Diatoms and Silicoflagellates

Some diatom species, especially those belonging to the genus *Chaetoceros*, have been known to cause mortalities of fish due to physical damage. In Scottish coastal waters, mortalities of farmed salmon have been linked to blooms of this organism. Examples include a bloom consisting predominantly of *C. wighami* in Loch Torridon,[99] as well as a mixed bloom of *C. debile* and the silicoflagellate, *Dictyocha speculum* (previously *Distephanus speculum*), in the Shetland Isles.[136] These were responsible for the deaths of farmed fish with a market value of several million pounds.

3.2.6 Other Species of Pelagic Microplankton

Other phytoplankton species present in the coastal waters of the UK may cause discolouration of the water but have little known direct human or ecosystem effects. For example, the coccolithophorid, *Emiliania huxleyi*, is known to bloom in the waters off Shetland,[137] as well as waters off the south west coast of England. One such incident, an extensive bloom which covered up to 16 000 km^2 at its peak, was reported in July 1999 off the south western coast of England.[138] The ciliate *Myrionecta rubra* (previously named *Mesodinium rubrum*) has frequently bloomed in Southampton Water[139] causing red discolouration of the surface water.

Blooms of the dinoflagellate, *Noctiluca scintillans*, have also been recorded in the Irish Sea in coastal waters of the Isle of Man (T. Shammon, personal communication), Northern Ireland (Food Standards Agency (NI), unpublished data) and the English Channel.[132] One of the most extensive was a bloom in 1982 that was recorded from Plymouth to the French coast and caused the water to look like 'tomato soup'.[132]

4 Safeguarding Health

4.1 Monitoring

Safeguarding human health from the effects of phytoplankton-derived toxins is most often achieved by monitoring, although a range of methodologies aimed at predicting the appearance of harmful algae are in development.

The approach to monitoring varies between countries and hence in this section, as an example, we describe the protocols adopted in Scottish and Northern Irish Waters within the UK. After the 1968 PSP outbreak in north east England, biotoxin monitoring was established in England, with a small monitoring programme introduced in Scotland at the same time. In Northern Ireland, regular testing of commercial shellfish beds was carried out for the presence of PSP toxins during the 1970s and 1980s (ref. 140). Following a series of negative results, testing ceased in the mid-1980s but was re-established again in 1990. However, in the mid-1990s, with the implementation of the EU Shellfish Hygiene Directive, monitoring for the presence of biotoxins in shellfish flesh expanded geographically and monitoring for the presence of toxin-producing species of phytoplankton was introduced.

Current monitoring programmes are driven by Food Hygiene regulations (EC) No 853/2004 and (EC) No 854/2004 that require member states to monitor toxin concentrations in shellfish flesh and the presence of marine biotoxin-producing phytoplankton. However, local differences in the implementation of these programmes exist, based on the risk of shellfish toxicity in the different locations as described in the examples below.

In Scotland there are c. 260 shellfish harvesting sites within 176 classified production areas. Shellfish from these areas are tested on a regular basis, as determined by seasonal risk, for the presence of biotoxins. As monitoring each individual site at a high enough temporal frequency to mitigate risk would be logistically and financially prohibitive, the production areas have been grouped into c. 90 'Pods', composed of a 'Representative Monitoring Point' (RMP) and 'Associated Harvesting Area/s' (AHA). RMPs are primarily chosen on a geographical basis, although factors such as toxin history, ease of access, and species present at the site are also taken into account. Shellfish samples for regulatory toxin testing are usually collected from RMPs only. Initially a risk assessment is carried out by the competent authority, the Food Standards Agency for Scotland. This is based on available information relating to the seasonal occurrence of different shellfish toxicity syndromes and the severity of the potential health effects. Based on this risk assessment, the sampling frequency applied at all sites is set as follows:

- PSP toxins: monitored weekly all year.
- DSP toxins: monitored weekly from April–November, fortnightly April–June, monthly January–March.
- ASP toxins: weekly from July–November, fortnightly April–June, monthly December–March.

If shellfish toxins are detected above action concentrations, shellfish harvesting is prohibited at the RMP and associated AHAs until toxin levels have reduced below safe regulatory thresholds. However, harvesting at an AHA associated with a closed RMP is possible if it can be demonstrated, for example through end-product testing, that toxins from shellfish collected from the AHA are below the regulatory limit.

In addition to the shellfish flesh toxicity testing, a representative group of about 37 RMP sites have also been selected for analysis and enumeration of the potentially harmful phytoplankton. Site selection is based on previous knowledge of bloom occurrence and geographic location. Phytoplankton monitoring occurs weekly over the period of April to September, with fortnightly sampling in March and October and monthly from November to February. The frequency of occurrence and spatial distribution of the three

Harmful Algal Blooms

major harmful phytoplankton genera in Scottish waters is displayed, for 2009, in Figure 4.

In Northern Ireland there are 44 classified shellfish harvesting beds in coastal waters. The approach to monitoring is similar to that in Scotland, the approximately 30 RMPs having been selected for routine sampling with sites chosen on the basis of a risk assessment (using data from sanitary surveys, and historical toxin and phytoplankton data). Positive biotoxin results from an RMP are used as a trigger for additional sampling at adjacent/associated

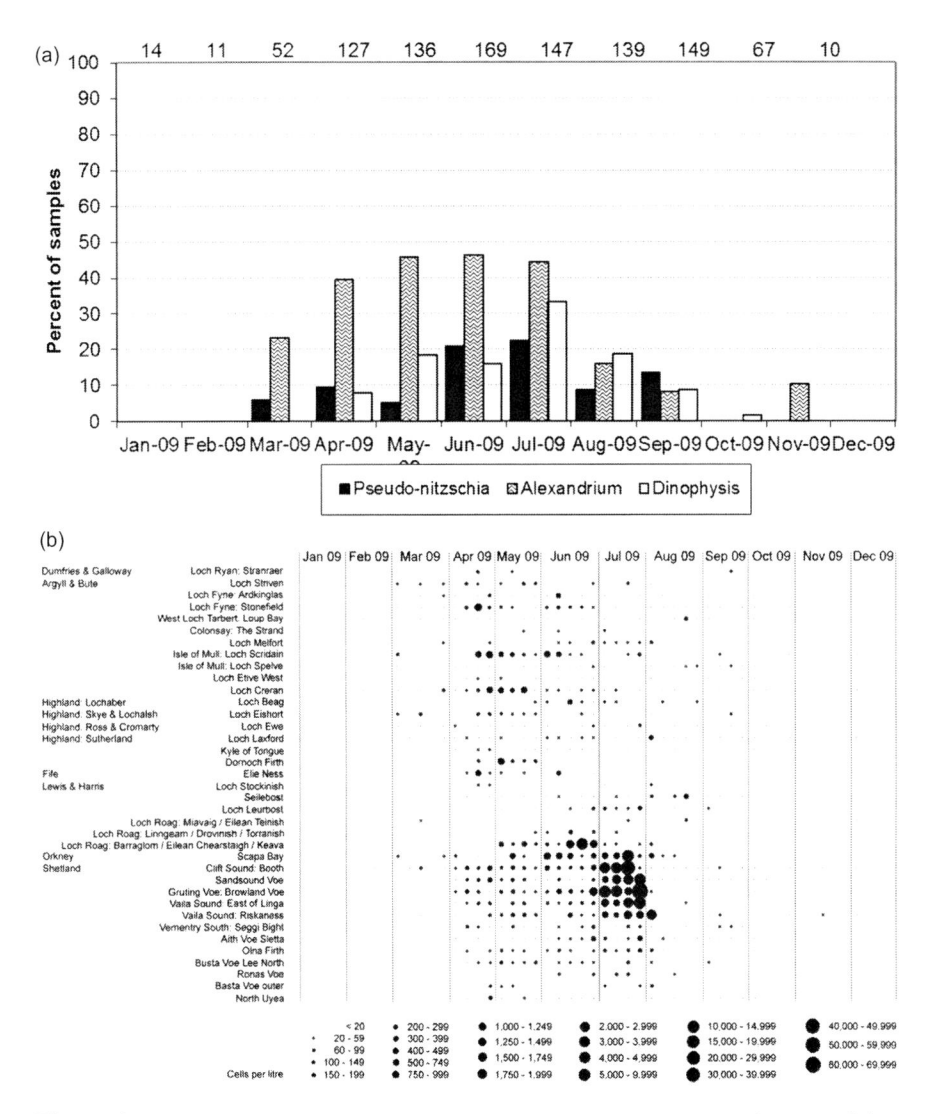

Figure 4 (a) Percentage of Scottish monitoring programme samples containing *Alexandrium*, *Pseudo-nitzschia* and *Dinophysis* during 2009. (b) Temporal and spatial appearance of *Alexandrium* in Scottish waters during 2009.

shellfish beds. In contrast to Scotland, the temporal frequency of the sampling regime is fortnightly for phytoplankton (all year round) and monthly for biotoxins (although sampling is rotated so that samples are received fortnightly from particular coastal areas).

As noted above, monitoring of shellfish toxins and causative phytoplankton is required by EU legislation. While data are limited, such an approach seems very successful in safeguarding human health, with relatively few recorded occurrences of shellfish poisoning in the UK on an annual basis. However, if there were no direct requirement on government to conduct such monitoring one might question whether it would occur, as the costs of monitoring are high and direct government control of food safety through monitoring is unusual. Standard food safety practice is for industry to take full responsibility for food safety based around Hazard Action Critical Control Points (HACCP). It is unclear whether the shellfish industry would be able to shoulder these costs, and many regions might not be able to sustain a shellfish industry and the positive benefits it brings to the economy. Moreover, a number of compelling reasons for monitoring exist. Firstly, it must be remembered that human illness generated by algal biotoxins is usually acute, sporadic and difficult to predict, making implementation of HACCP problematic. Furthermore, shellfish are usually eaten whole, with limited processing. Finally, cooking, even if it does occur, has little or no influence on toxin content.

4.2 Are Algal Toxins a Public Health Problem?

Worldwide, the health risks from algal toxins are clear, although estimates of the magnitude of the problem vary. Wang[141] suggested that algal toxins result in 50 000–500 000 intoxication incidents per year, with an overall mortality rate of 1.5% on a global basis. The lower end of Wang's suggested range is relatively consistent with Hallegraeff[27] who reported worldwide that c. 300 people a year die as a result of HAB-related events. Within these illnesses, Ciguatera Fish Poisoning (CFP) is the most frequently reported seafood-toxin syndrome in the world, with substantial physical and functional impact for sufferers. It is estimated that 10 000–50 000 people per year suffer from CFP, although the true incidence is difficult to ascertain due to under-reporting. Friedman et al.[18] suggested that only 2–10% of CFP cases are reported to authorities.

The economic impact of algal toxins is also clear, with Hoagland et al.[142] estimating that for the United States the average annual cost (1987–1992) of HABs (in terms of the effects on public health, commercial fisheries and recreation/tourism) was US$50 million.

In the UK there is formal national guidance to the National Health Service and local authorities on cyanobacteria in inland and inshore waters. Each region has an annually updated action plan that includes protocols for testing, thresholds that trigger action, notification, publicity and retesting. The action to be taken depends on the location of the cyanobacteria, its abundance and the use of the watercourse. The aim of the plan is to reduce skin contact and avoid ingestion. Signs are posted advising the public to avoid swimming, sailing or eating fish caught in areas with abundant growth. Public water supply has a threshold for maximum permitted concentration of toxin in drinking water and there is also consideration of special cases such as renal dialysis, at home and in hospitals.

In contrast, for shellfish toxins no formal medical response plan exists. This is partly because it is thought that incidences of shellfish poisoning are relatively rare, with the primary monitoring programmes described above preventing most illnesses, at least with regard to their acute presentation due to high dose exposure. However, as there is no formal system of recording incidences of shellfish toxicity in humans, the scale of illness resulting from shellfish toxins is poorly quantified and it is clear that many cases will go unreported. Furthermore, confirmatory diagnostic testing only exists for the contaminated transvector (which is not readily available to most healthcare providers), and not for human fluids (*i.e.* no biomarker exists). This makes diagnosis of even those suspected cases that are reported unlikely and, if actually diagnosed, established treatment is only supportive.

Almost nothing is known about possible chronic health effects after high-dose, acute exposure or after long-term, low-level exposure; the latter is of concern, given the carcinogenicity of some of these compounds (*e.g.* microcystin in drinking water). In addition, little is known about the acute or chronic effects of these toxins in potentially susceptible sub-populations (children, the elderly and persons with chronic neurologic diseases). In general, outreach and education about the HAB diseases in humans targeted at healthcare providers, public health officials and resource managers is almost non-existent.

With newly emerging human illnesses associated with exposure to HAB toxins, including aerosolised toxins, brevetoxin fish poisoning, and azaspiracid shellfish poisoning, as well as the apparent increase in HABs in some parts of the world, the development of appropriate diagnostic tools in humans and other animals for other HAB-derived illnesses is essential. These can be used to establish surveillance baselines through appropriate diagnosis and reporting, and to explore issues of appropriate treatment and chronic health effects. In addition, HAB outreach and education targeted at healthcare providers, public health officials and resource managers, as well as at vulnerable populations such as tourists, would assist in the primary (*e.g.* completely preventing exposure through monitoring), secondary (*e.g.* early diagnosis), and tertiary (*e.g.* appropriate treatment) prevention of these illnesses.

4.3 Early Warning Methodologies and Mitigation

Despite extensive monitoring of harmful phytoplankton and biotoxins, there is, with a few exceptions, little robust understanding of where and when blooms and toxicity events will occur, their magnitude, and the level of toxicity. Few systems are in place to identify high risk areas or to estimate or predict the risk of toxic events, as to make such risk assessments it is necessary to understand the complex relationship between environmental factors, HABs and shellfish poisoning.

The most direct early warning of the potential for HAB-related events is the presence of the phytoplankton themselves as, particularly for shellfish toxicity events, there is often a one or two week lag between an increase in phytoplankton density and significant (above regulatory limit) toxicity in the shellfish. Due to environmental control of toxin content in cells, variable filtration rates and, in part, because of phytoplankton sampling issues (related to frequency of sampling and not sampling at the same time or tidal state), relationships between harmful phytoplankton abundance and shellfish toxicity are not always clear cut. However, detailed studies[143] have shown phytoplankton monitoring to provide valuable early warning. An example of increasing harmful phytoplankton prior to a shellfish toxicity event from Scottish waters is presented in Figure 5.

The approach to the use of such phytoplankton-based early warning is variable. In the Scotland and Northern Ireland, regulatory decisions are not made on the basis of phytoplankton data. Rather, the information is made freely available on the internet to inform harvesting decisions. However, in other countries (for example, France) shellfish flesh monitoring is actively targeted based on elevated counts of potentially harmful phytoplankton.

In some regions of the world HAB forecasting systems have been developed with the specific aim of safeguarding human health. Perhaps the most sophisticated operational system is the NOAA (National Ocean and Atmospheric

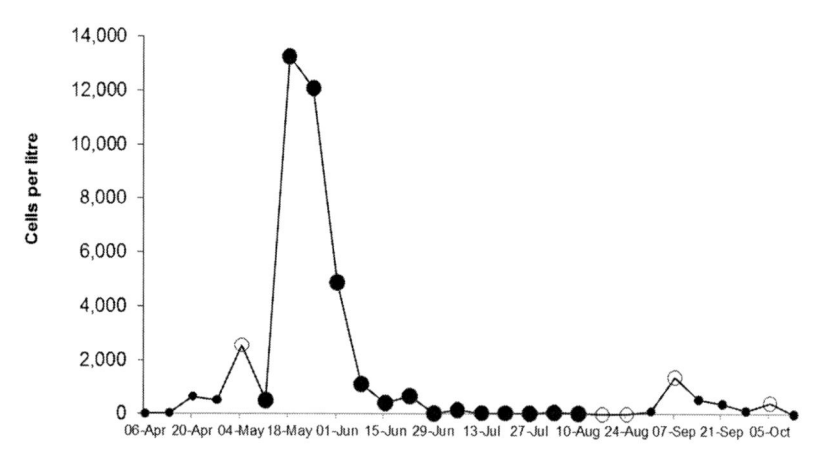

Figure 5 Development of shellfish toxicity following a *Dinophysis* bloom in Loch Fyne on the Scottish West Coast in 2009. Cell density increases to above regulatory threshold levels on the 20th of April, but toxin levels only reach values that result in site closure during mid May. (Small symbols: no shellfish toxicity, large open symbols: clinical signs of shellfish toxicity, large closed symbols: site closed due to DSP).

Administration) Harmful Algal Bloom Operational Forecast (HAB-OFS) that operates in Florida and Texas in the US.[144] The success of the system relies on the use of a number of complementary methodologies including satellite imagery, field observations, mathematical models, public health reports and data from instrumented moorings to predict HAB blooms and their movement. A key to the success of such forecasting systems is that the biology of the HAB species of concern is closely coupled to a relatively simple (and therefore predictable) hydrodynamic regime.

While such systems are potentially of great value they are expensive to operate. Moreover, a major constraint to the wider development of HAB prediction systems elsewhere is that we simply do not know enough about the ecophysiology of many of the key species to be able to predict their occurrence. Furthermore, since toxicity can be induced by as few as 100 cells l^{-1}, densities that are typical of the normal assemblage of species found in coastal waters in summer, modelling the occurrence of HAB species, rather than similar but benign organisms, is a major challenge.

4.4 Introductions and Transfers of New Species

The wealth of new observations suggests that HAB-related problems are increasing. Whether this is a true increase or rather a result of our increased vigilance and monitoring activity is unclear.[17] However, a number of factors can influence and potentially increase the spatial and temporal appearance of HABs and their health consequences.

The potential for human activity to inadvertently transfer a range of plants and animals from one region of the world to another is well documented. A number of examples of this phenomenon have been demonstrated for HAB species with potentially unfortunate consequences; selected examples are provided below.

Hallegraeff and Bolch[145] investigated the presence of dinoflagellate cysts in the sediment found in cargo vessel ballast tanks entering Australian ports and found that, of the sediment samples collected from 80 vessels, 40% contained viable dinoflagellate cysts of non-toxic species and 6% contained the cysts of the toxin-producing species, *Alexandrium catenella* and *Alexandrium tamarense*. A further study surveyed 343 vessels entering Australian ports and found that more than 200 of the vessels contained sediment in the bottom of their ballast tanks and of these 50% contained dinoflagellate cysts.[146]

A similar study undertaken in Scotland[147] gave comparable results. Of 127 vessels that were in ballast entering Scottish ports, motile dinoflagellate cells were found in the ballast water of 76% of these, and cysts were found in 61% of 92 sediment samples collected.

These studies clearly show that motile cells and resting cysts can be transported in ballast water and sediment, with Hallegraeff and Bolch[145] concluding that the evidence pointed to the 'distinct possibility' that *Gymnodinium catenatum* had been introduced to coastal waters around Hobart in Tasmania (Australia).

4.5 Climate Change

A major review of HAB-specific issues has recently been published by Hallegraeff[148] with the impact in UK waters in particular also being recently discussed.[149] The role of climate change on seafood safety has also recently been reviewed.[150] Hence, this complicated issue is dealt with only briefly below.

Part of the difficulty in forecasting the effect of climate change on harmful algal blooms arises because we do not properly understand why some species of micro-algae produce biotoxins, or whether the seeming increase in HABs in recent decades is real and, if it is, what has caused it. In contrast, there is very little doubt that our planet is getting warmer and will continue to warm, because of the physical consequences of increased atmospheric greenhouse gases. And warmer, on average, means wetter on average, because warmer air will carry more water vapour. The difficulties return, however, when we ask how much warmer and wetter will it be in, say, 50 years time in northern Britain, and how will this influence physical conditions in the sea and nutrient discharges from land to sea.

Making the broadest generalisation, pelagic diatoms are an ice-age lifeform, flourishing in cool, nutrient-rich, turbulent waters. In contrast, dinoflagellates are adapted to warm and stratified seas. Thus, in these broad terms, we might expect the pelagic balance to shift from diatoms towards dinoflagellates as the world warms, giving greater likelihood of HABs. Some evidence for this is provided by studies of sediment records. As discussed above, the PSP-causative organism, *Gymnodinium catenatum*, is characteristic of warm waters, causing problems, for example, in the Rias of northern Spain[151] but not at higher latitudes. However, sediment core studies from Scandinavian waters suggest that this dinoflagellate was once abundant in coastal waters as far north as Bergen in Norway during warm periods in the last two millennia.[152] Such results would suggest the expansion of warm-water species at the expense of cold-water species and hence the potential for range expansion of some HAB organisms.

However, there are complications. Postglacial changes in sea-level have altered tidal mixing regimes, and this can cause shifts in the location of dinoflagellate growth zones,[153] as can changes in freshwater discharge. Anthropogenic enrichment of nutrients in coastal waters has doubtless altered the diatom–dinoflagellate balance during the last centuries, and there is some evidence for this in cyst records.[154] As much time-series data for phytoplankton are limited to a few decades (at best), some scientists have attempted to relate sub-decadal fluctuations in plankton to year-to-year changes in weather patterns associated with sub-global climatic systems such as the North Atlantic Oscillation (NAO) and the El Niño Southern Oscillation (ENSO). For example, Breton *et al.*[155] reported that "Statistical analysis of 14 yr (1988–2001) of intensive phytoplankton monitoring ... in the ... Southern Bight of the North Sea ... indicates that the long-term diatom biomass trend and the spring dominance of *Phaeocystis* colonies over diatoms are determined by the combined effect of the North Atlantic Oscillation (NAO) and freshwater and continental nitrate carried by the Scheldt [river]." Where such correlations have been demonstrated, it is tempting to use them to forecast – in this case, increased abundance of *Phaeocystis* as a result of warmer and wetter conditions (corresponding to the positive phase of the NAO, meaning a stronger atmospheric pressure gradient from the Azores to Iceland). However, even if such relationships are robust they are likely to be species- and location- specific. For example, Gowen *et al.*[17] could find no relationship between large fluctuations in HABs in south China coastal waters[156] and the ENSO.

Acknowledgements

Some of the material presented in this chapter was based on the Defra-funded study 'Anthropogenic nutrient enrichment and blooms of harmful micro-algae', project ME2208. The work also draws on discussions at a workshop funded by the UK Joint Environment and Human Health Programme (NERC, EA, Defra, MOD, MRC, The Wellcome Trust, ESRC, BBSRC, EPSRC and HPA). We thank all the participants. During chapter preparation, Keith Davidson was in receipt of funding from the NERC Oceans 2025 programme and the EU Interreg IV Northern Periphery programme 'WATER'. Finally, we thank Sarah Swan and Bhavani Narayanaswamy for assistance with figure preparation.

References

1. P. G. Falkowski, R. T. Barber and V. Smetacek, *Science*, 1998, **281**, 200.
2. P. Tett, in *Light and Life in the Sea*, ed. P. J. Herring, A. K. Campbell, M. Whitfield and L. Maddock, Cambridge Univ. Press, Plymouth, 1990, 59.
3. H. U. Svedrup, *J. Cons. Perm. Int. Explor. Mer.*, 1953, **18**, 287.
4. V. Smetacek and U. Passow, *Limnol Oceanogr.*, 1990, **35**, 228.
5. D. L. Tang, H. Kawamura, T. Van Dien and M. Lee, *Mar. Ecol:. Prog. Ser.*, 2004, **268**, 31.
6. Y. Azov, *Mar. Pollut. Bull.*, 1991, **23**, 225.
7. R. L. Smith, *Oceanogr. Mar. Biol. Ann. Rev.*, 1968, **6**, 11.
8. S. W. Nixon, *Ophelia*, 1995, **41**, 199.
9. R. J. Gowen, P. Tett, K. Kennington, D. K. Mills, T. M. Shammon, *et al.*, *Estuarine Coastal Shelf Sci.*, 2008, **76**, 239.
10. A. Sournia, in *Harmful Marine Algal Blooms*, ed. P. Lassus, G. Arzul, E. Erard-Le Denn, P. Gentien and C. Marcaillou-Le Baut, Lavoisier Publishing, Paris, 1995, 103.
11. S. M. Marshall and A. P. Orr, *J. Mar. Biol. Assoc. UK*, 1930, **16**, 853.
12. R. J. Gowen, B. M. Stewart, D. K. Mills and P. Elliott, *J. Plankton Res.*, 1995, **17**, 753.
13. T. Okaichi, in *Sustainable Development in the Seto Inland Sea, Japan - From the Viewpoint of Fisheries*, ed. T. Okaichi and T. Yanagi, Terra Scientific Publishing Company, Tokyo, 1997, 251.

14. M. Parker and P. Tett, *Special Meeting on the Causes, Dynamics and Effects of Exceptional Marine Blooms and Related Events Rapport et Proces-verbaux des Reunions*, Conseil International pour l'Exploration de la Mer, 1987, 187.
15. T. J. Smayda, *Limnol. Oceanogr.*, 1997, **42**, 1137.
16. P. Tett, R. Gowen, D. Mills, T. Fernandes, L. C. Gilpin, *et al.*, *Mar. Pollut. Bull.*, 2007, **55**, 282.
17. R. Gowen, P. Tett, E. Bresnan, K. Davidson, A. Gordon, A. McKinney, S. Milligan, D. Mills, J. Silke and A. M. Crooks, *Anthropogenic Nutrient Rnrichment and Blooms of Harmful Micro-algae*, DEFRA, 2009.
18. M. A. Friedman, L. E. Fleming, M. Fernandez, P. Bienfang, K. Schrank *et al.*, *Marine Drugs*, 2008, **6**, 456.
19. H. A. Anderson and M. S. Wolff, *Environ. Res*, 2005, **97**, 125.
20. K. Davidson and E. Bresnan, *Environ. Health*, 2009, **8**, S12.
21. DEFRA, aquaculture, http://www.archive.defra.gov.uk/foodfarm/fisheries/farm-health/aquaculture.htm, 2010 (last accessed 28/12/10).
22. T. J. Smayda, *Relationship to Fish Farming and Regional Comparisons - A Review*, Scottish Executive Environment Group Paper 2006/3, 2005, http://www.scotland.gov.uk/Resource/Doc/92174/0022031.pdf (last accessed 28/12/10).
23. A. D. Cembella, in *The Physiological Ecology of Harmful Algal Blooms*, ed. D. M. Anderson, A. D. Cembella and G. M. Hallegraeff, Springer-Verlag, Heidelberg, 1998, 381.
24. J. Landsberg, *Rev. Fish. Sci.*, 2002, **10**, 113.
25. J. R. Deeds, J. H. Landsberg, S. M. Etheridge, G. C. Pitcher and S. W. Longan, *Mar. Drugs*, 2008, **6**, 308.
26. J. F. R. Medcof, in *Toxic Dinoflagellates*, ed. A. M. Anderson, A. W. While and D. G. Baden, Elsevier, Amsterdam, 1985, 1.
27. G. Hallegraeff, *Phycologia*, 1993, **32**, 79.
28. R. V. Azanza and F. J. R. Taylor, *Ambio*, 2001, **30**, 356.
29. J. L. MacLean, in *Biology, Epidemiology and Management of Pyrodinium Red Tides*, ICLARM Conference proceedings, ed. G. M. Hallegraeff and J. L. MacLean, 1989, 1.
30. D. M. Anderson, in *The Physiological Ecology of Harmful Algal Blooms*, ed. D. M. Anderson, A. D. Cembella and G. M. Hallegraeff, Springer-Verlag, Heidelberg, 1998, 29.
31. E. L. Lilly, K. M. Halaynch and D. M. Anderson, *J. Phycol.*, 2007, **43**, 1329.
32. H. Sommer and K. F. Meyer, *Arch. Pathol.*, 1937, **24**, 560.
33. S. Fraga, B Reguera and I. Bravo, in *Toxic Marine Phytoplankton*, ed. E. Granéli, B. Sundström, L. Edler and D. M. Anderson, Elsevier, New York, 1990, 149.
34. B. Dale, B. A. Madsen, K. Nordberg and T. A. Thorsen, in *Toxic Phytoplankton Blooms in the Sea*, ed. T. J. Smayda and T. Shimizu, Elsevier, Amsterdam, 1993, 47.

35. S. S. Bates, D. L. Garrison and R. A. Horne, in *The Physiological Ecology of Harmful Algal Blooms*, ed. D. M. Anderson, A. D. Cembella and G. M. Hallegraeff, Springer-Verlag, Heidelberg, 1998, 267.
36. S. S. Bates, *J. Phycol.*, 2000, **36**, 978.
37. S. S. Bates, J. C. Bird, A. S. W. de Freitas, R. Foxall, M. Gilgan, *et al.*, *Can. J. Fish. Aquat. Sci.*, 1989, **46**, 1203.
38. C. A. Scholin, F. Gulland, G. J. Doucette, S. Benson, M. Busman, *et al.*, *Nature*, 2000, **403**, 80.
39. J. L. Martin, K. Haya and D. J. Wildish, in *Toxic Phytoplankton Blooms in the Sea*, ed. T. J. Smayda and Y. Shimizu, Elsevier, Amsterdam, 1993, 613.
40. J. Y. Couture, M. Levasseur, E. Bonneau, C. Desjardins, G. Sauve, *et al.*, *Can. Tech. Rep. Fish. Aquat. Sci./Rapp. Tech. Can. Sci. Halieut. Aquat.*, 2375, 2001.
41. S. S. Bates, C. Léger, J. M. White, N. MacNair, J. M. Ehrman, M. Levasseur and G. Sauvé, *Proceedings of the 17th International Diatom Symposium*, ed. M. Poulton, Biopress, Bristol, 2002, 7.
42. Y. L. Pan, D. V. S. Rao, K. H. Mann, R. G. Brown and R. Pocklington, *Mar. Ecol.: Prog. Ser.*, 1996, **131**, 225.
43. J. Fehling, K. Davidson, C. J. Bolch and S. S. Bates, *J. Phycol*, 2004, **40**, 674.
44. E. Rue and K. Bruland, *Mar. Chem.*, 2001, **76**, 127.
45. M. Kat, *Antonie van Leeuvehhoek*, 1983, **49**, 417.
46. T. Yasumoto, Y. Oshima, W. Sugawara, Y. Fukuyo, H. Oguri, T. Igarashi and N. Fujita, *Bull. Japanese Soc. Sci. Fish.*, 1980, **46**, 1405.
47. M. A. Quilliam and J. L. Wright, in *Manual on Harmful Marine Microalgae*, ed. G. M. Hallegraeff, D. M. Anderson and A. D. Cembella, IOC manuals and guides, Unesco, 1995, **33**, 95.
48. E. Dahl, A. Rogstad, T. Aune, V. Hormazabal and B. Underdal, in *Marine Algal Blooms*, ed. P. Lassus, G. Arzul, E. Erard-Le Denn, P. Gentien and C. Marcaillou-Le Baut, Lavoisier publishing, Paris, 1995, 738.
49. R. Raine, G. McDermott, J. Silke, K. Lyons, G. Nolan and C. Cusack, *J. Mar. Syst.*, 2010, **83**, 150.
50. C. Belin, in *Toxic Phytoplankton Blooms in the Sea*, ed. T. J. Smayda and Y. Shimizu, Elsevier, Amsterdam, 1993, 469.
51. S. Fraga and F. J. Sanches, in *Toxic Dinoflagellates*, ed. A. M. Anderson, A. W. While and D. G. Baden, Elsevier, Amsterdam, 1985, 51.
52. T. McMahon, E. Nixon and J. Silke, *Harmful Algae News*, 1995, **10/11**, 6.
53. T. McMahon, R. Raine and J. Silke, in *Proceedings of the 8th International Conference on Harmful Algae*, ed. B. Reguera, J. Blanco, M. L. Fernandez and T. Wyatt, IOC of UNESCO, 1998, 128.
54. M. G. Park, S. Kim, H. S. Kim, G. Myung and Y. G. Kang, *Aquat. Microb. Ecol.*, 2006, **45**, 101.
55. Y. Pan, A. D. Cembella and M. A. Quillam, *Mar. Biol.*, 1999, **134**, 541.
56. L. A. Stobo, J.-P. Lacaze, A. C. Scott, J. Petrie and E. A. Turrell, *Toxicon*, 2008, **51**, 635.

57. B. Paz, A. H. Daranas, M. Norte, P. Riobó, J. Franco and J. J. Fernández, *Mar. Drugs*, 2008, **6**, 73.
58. T. McMahon and J. Silke, *Harmful Algal News*, 1996, **14**, 1.
59. E. Ito, M. Satake, K. Ofuji, N. Kurita, T. McMahon, K. James and T. Yasumoto, *Toxicon*, 2000, **38**, 917.
60. M. J. Twiner, N. Rehmann, P. Hess and G. J. Doucette, *Mar. Drugs*, 2008, **6**, 39.
61. K. Ofuji, M. Satake, T. McMahon, J. Silke, K. J. James, H. Naoki, Y. Oshima and T. Yasumoto, *Nat. Toxins*, 1999, **7**, 99.
62. U. Tillmann, M. Elbrachter, B. Krock, U. John and A. D. Cembella, *Eur. J. Phycol.*, 2009, **44**, 63.
63. S. M. Watkins, A. Reich, L. E. Fleming and R. Hammond, *Mar. Drugs*, 2008, **6**, 431.
64. F. M. van Dolan, *Environ. Health Perspect.*, 2000, **108**(Suppl 1), 133.
65. H. A. Magaña, C. Contreras and T. A. Villareal, *Harmful Algae*, 2003, **2**, 163.
66. M. Monti, M. Minocci and L. Beran Aiveša, *Mar. Pollut. Bull*, 2008, **54**, 598.
67. C. Brescianini, C. Grillo, N. Melchiorre, R. Bertolotto, A. Ferrari, *et al.*, *Eur. Surveillance*, 2006, **11**, E060907.3.
68. P. Ciminiello, C. Dell'Aversano, E. Fattorusso, *et al.*, *Anal. Chem.*, 2006, **78**, 6153.
69. G. Asaeda, *J. Energ. Med.*, 2001, **20**, 263.
70. L. Lehane and R. J. Lewis, *Int. J. Food Microbiol*, 2000, **61**, 91.
71. G. M. Nicholson and R. J. Lewis, *Mar. Drugs*, 2006, **4**, 88.
72. L. Lehane, *Med. J. Aust.*, 2000, **174**, 174.
73. C. Dow and U. Swoboda, in *The Ecology of Cyanobacteria*, ed. B. A. Whitton and M. Potts, Kluwer Academic Publishers, Boston, MA, 2000, 614.
74. N. J. T. Osborne, P. M. Webb and G. R. Shaw, *Environ. Int.*, 2001, **27**, 381.
75. I. Chorus and J. Bartram, *Toxic Cyanobacteria in Water: A Guide to their Public Health Consequences, Monitoring and Management*, St Edmundsbury Press, Bury St Edmunds, 1999.
76. K. G. Sellner, *Limnol. Ocenaogr.*, 1997, **42**, 1089.
77. S. Kaartvedt, T. M. Johnsen, D. L. Aksnes, U. Lie and H. Svendsen, *Can. J. Fish. Aquat. Sci.*, 1991, **48**, 2316.
78. T. Okaichi, in *Red Tides Biology, Environmental Science, and Toxicology*, ed. T. Okaichi, D. M. Anderson and T. Nemoto, Elsevier, Amsterdam, 1989, 137.
79. S. Yamochi, in *Red Tides Biology, Environmental Science, and Toxicology*, ed. T. Okaichi, D. M. Anderson and T. Nemoto, Elsevier, Amsterdam, 1989, 253.
80. E. A. Black, J. N. C. Whyte, J. W. Bagshaw and N. G. Ginther, *J. Appl. Ichthyol.*, 1991, **7**, 168.
81. L. MacKenzie, *J. Appl. Phycol.*, 1991, **3**, 19–34.
82. A. T. Davidson and H. J. Marchant, in *Progress in Phycological Research*, ed. F. E. Round and D. J. Chapman, Biopress, Bristol, 1992, **8**, 1.
83. C. Lancelot, in *The Science of the Total Environment*, Elsevier, Amsterdam, 1995, **165**, 83.

84. P. S. Liss, G. Malin, S. M. Turner and P. M. Holligan, *J. Mar. Syst.*, 1994, **5**, 41.
85. R. E. Savage, in *Fishery Investigations Series II*, HM Stationary Office, London, 1930, vol. 12.
86. F. H. Chang, *New Zealand. J. Mar. Freshwater Res.*, 1983, **17**, 165.
87. Y. Z. Qi, J. F. Chen, Z. H. Wang, N. Xu, Y. Wang, P. P. Shen, S. H. Lu and I. J. Hodgkiss, *Hydrobiol.*, 2004, **512**, 209.
88. R. T. Aanesen, H. C. Eilertsen and O. B. Stabell, *Toxicology*, 1998, **40**, 109.
89. O. B. Stabell, R. T. Aanesen and H. C. Eilertsen, *Aquat. Toxicology*, 1999, **44**, 279.
90. H. Zhang, W. Litaker, M. W. Vandersea, P. Tester and S. Lin, *J. Plankton. Res.*, 2008, **30**, 905.
91. D. I. Kim, Y. Matsuyama, S. Nagasoe, M. Yamaguchi, Y. H. Yoon, Y. Oshima, N. Imada and T. Honjo, *J. Plankton Res.*, 2004, **26**, 61.
92. M. Vargas-Montero, E. Freer, R. Jimenez-Montealegre and J. C. Guzman, *African J. Mar. Sci.*, 2006, **28**, 215.
93. H. K. Kim, *Ocean Research (Seoul)*, 1997, **19**, 185.
94. E. Dahl and K. Tangen, in *Toxic Phytoplankton blooms in the Sea*, ed. T. J. Smayda and Y. Shimizu, Elsevier, Amsterdam, 1993, **3**, 15.
95. R. Raine, B. Joyce, J. Richard, Y. Pazos, M. Moloney, K. J. Jones and J. W. Patching, *ICES J. Mar. Sci.*, 1993, **50**, 461.
96. K. Davidson, P. Miller, T. A. Wilding, J. Shutler, E. Bresnan, K. Kennington and S. Swan, *Harmful Algae*, 2009, **8**, 349.
97. R. J. Roberts, A. Bullock, M. Turner, K. J. Jones and P. Tett, *J. Mar. Biol. Assoc. UK*, 1983, **63**, 741.
98. M. Satake, Y. Tanaka, Y. Ishikura, Y. Oshima, H. Naoki and T. Yasumoto, *Tetrahedron Lett.*, 2005, **46**, 3537.
99. D. W. Bruno, G. Dear and D. D. Seaton, *Aquaculture*, 1989, **78**, 217.
100. P. Cleve, in *The Seasonal Distribution of Atlantic Plankton Organisms*, ed. D. F. Bonniers, Göteborg, 1900.
101. W. A. Herdman and W. Riddell, *Proc. Trans. Liverpool Biol. Soc.*, 1911, **25**, 132.
102. W. A. Herdman and W. Riddell, *Proc. Trans. Liverpool Biol. Soc.*, 1912, **26**, 225.
103. M. V. Lebour, *J. Mar. Biol. Assoc. U.K.*, 1917, **11**, 132.
104. M. V. Lebour, *The Dinoflagellates of Northern Seas*, Marine Biological Association of Plymouth, 1925.
105. C. Collins, J. Graham, L. Brown, E. Bresnan, J.-P. Lacaze and E. A. Turrell, *J. Phycol.*, 2009, **45**, 692.
106. N. Touzet, K. Davidson, R. Pete, K. Flanagan, G. R. McCoy, Z. Amzil, M. Maher, A. Chapelle and R. Raine, *Protist*, 2010, **161**, 370.
107. N. Touzet, H. Farrell, A. Ní Rathaille, P. Rodriguez, A. Alfonso, L. M. Botana and R. Raine, *Deep Sea Res.*, 2010, **57**, 268.
108. T. Wyatt and F. Saborido-Rey, in *Toxic Phytoplankton Blooms in the Sea*, ed. T. J. Smayda and T. Shimizu, Elsevier, Anmsterdam, 1993, 73.

109. P. Tett and V. Edwards, *Review of Harmful Algal Blooms in Scottish Coastal Waters*, SEPA, Edinburgh, 2002.
110. P. A. Ayres, D. D. Seaton and P. Tett, *ICES*, CM 1982/L:38.
111. I. Joint, J. Lewis, J. Aiken, R. Proctor, G. Moore, W. Higman and M. Donald, *J. Plankton. Res.*, 1997, **19**, 937.
112. A. W. Berry, *Harmful Algae News*, 1997, **16**, 8.
113. J. S. Combe, *Edinburgh Med. Surg. J.*, 1828, **29**, 86.
114. P. Ayres, *Environ. Health*, 1975, **83**, 261.
115. P. A. Ayres and M. Cullem, *Paralytic Shell Fish Poisoning: An Account on Investigations into Mussel Toxicity in England 1968–1977*, MAFF, Directorate of Fisheries, London, 1978.
116. G. A. Robinson, *Nature*, 1968, **220**, 22.
117. J. A. Adams, D. D. Seaton, J. B. Buchanan and M. R. Longbottom, *Nature*, 1968, **220**, 24.
118. J. C. Coulson, G. R. Potts, I. R. Deans and S. M. Fraser, *Nature*, 1968, **220**, 23.
119. ICES C.M.1991/Poll:3, *Report of the Working Group on Phytoplankton and the Management of their Effects*, ICES, 1991.
120. E. Bresnan, K. Davidson, R. Gowen, W. Higman, L. Lawton, J. Lewis, L. Percy, A. McKinney, S. Milligan, T. Shammon, S. Swan, in *Relating Harmful Phytoplankton to Shellfish Poisoning and Human Health*, ed. K. Davidson and E. Bresnan, 2008, 11, http://www.sams.ac.uk/research/departments/microbial-molecular/news/hab-workshop-report-published/ (last accessed, 28/12/2010).
121. J. Fehling, K. Davidson, C. J. Bolch and P. Tett, *Mar. Ecol.: Prog. Ser.*, 2006, **323**, 91.
122. J. Fehling, D. H. Green, K. Davidson, C. J. Bolch and S. S. Bates, *J. Phycol.*, 2004, **40**, 622.
123. F. T. O. Krogstad, W. C. Griffith, E. M. Vigoren and E. M. Faustman, *J. Appl. Phycol.*, 2009, **21**, 745.
124. ICES 2008/OCC:03 IC, *Report of the ICES-IOC Working Group on Harmful Algal Bloom Dynamics*, ICES, Galway, 2008.
125. S. M. Nascimento, D. A. Purdie and S. Morris, *Toxicon*, 2005, **45**, 633.
126. S. Morris, B. Stubbs, C. Brunet and K. Davidson, *Spatial Distributions and Temporal Profiles of Harmful Phytoplankton, and Lipophilic Toxins in Common Mussels and Pacific Oysters from Four Scottish Shellfish Production areas Areas*, Food Standards Agency, Scotland, 2010.
127. ICES 2002/C:03 IC, *Report of the ICES-IOC Working Group on Harmful Algal Bloom Dynamics*, ICES, Bermuda, 2002.
128. J. Orton, *Nature*, 1923, **111**, 773.
129. P. Jones and S. Haq, *J. Cons. Perm. Int. Explorat. Mer.*, 1963, **28**, 8.
130. ICES C.M.1993/ENV:7, *Report of the Working Group on Phytoplankton and the Management of their Effects*, ICES, Copenhagen, 1993.
131. A. G. Davies, I. Demadariaga, B. Bautista, F. Fernandez, D. S. Harbour, P. Serret and P. R. G. Tranter, *J. Mar. Biol. Assoc. UK*, 1992, **72**, 708.

132. G. T. Boalch, *Rapports et Proces-Verbaux des Reunions Conseil International pour l'Exploration de la Mer*, 1987, **187**, 94.
133. ICES 2006/OCC:04 IC, *Report of the ICES-IOC Working Group on Harmful Algal Bloom Dynamics*, ICES, Gdynia, 2006.
134. P. Tett, *Phytoplankton and the Fish Kills in Loch Striven*, Scottish Marine Biological Association Internal Report, 1980.
135. R. J. Gowen, J. Lewis and A. M. Bullock, *A Flagellate Bloom and Associated Mortalities of Farmed Salmon and Trout in Upper Loch Fyne*, Scottish Marine Biological Association Internal Report, 1982, No.17.
136. J. W. Treasurer, F. Hannah and D. Cox, *Aquaculture*, 2003, **218**, 103.
137. R. N. Head, D. W. Crawford, J. K. Egge, R. P. Harris, S. Kristiansen *et al.*, *J. Sea Res.*, 1998, **39**, 255.
138. T. J. Smyth, G. F. Moore, S. B. Groom, P. E. Land and T. Tyrrell, *Appl. Optics*, 2002, **41**, 7679.
139. D. W. Crawford, D. A. Purdie, A. P. M. Lockwood and P. Weissman, *Estuarine Coastal Shelf. Sci.*, 1997, **45**, 799.
140. W. J. McCaughey and J. N. Campbell, *Harmful Algae News*, 1992, **3**, 3.
141. D. Z. Wang, *Mar. Drugs*, 2008, **6**, 349.
142. P. Hoagland, D. M. Anderson, Y. Kaoru and A. W. White, *Estuaries*, 2002, **25**, 819.
143. K. Davidson, J. McElhiney, L. Murray and M. Algoet M, in *Scotland's Seas: Towards Understanding their State*, ed. J. M. Baxter, I. L. Boyd, M. Cox, L. Cunningham, P. Homes and C. F. Moffat, Fisheries Research Services, Aberdeen, 2008, 126.
144. R. P. Stumpf, in *Proceedings XII International Conference on Harmful Algae. IOC of Unesco*, ed. Ø. Moestrup, Copenhagen, 2008, 96.
145. G. Hallegraeff and C. J. Bolch, *Mar. Pollut. Bull.*, 1991, **22**, 27.
146. G. M. Hallegraeff and C. J. Bolch, *J. Plankton. Res.*, 1992, **14**, 1067.
147. E. M. MacDonald and R. D. Davidson, in *Proceedings of the 8th International Conference on Harmful Algae*, ed. B. Reguera, J. Blanco, M. L. Fernandez and T. Wyatt, IOC of UNESCO, 1998, 220.
148. G. Hallegraeff, *J. Phycol.*, 1993, **46**, 220.
149. E. Bresnan, L. Fernand, K. Davidson, M. Edwards, S. Milligan, R. Gowen, J. Silke, S. Kröger and R. Raine, in MCCIP Annual Report Card 2010–11, MCCIP Science Review, http://www.mccip.org.uk/arc (last accessed 2/1/2011).
150. A. Marques, M. L. Nunes. S. K. Moore and M. S. Strom, *Food Res. Int.*, 2010, **43**, 1766.
151. F. G. Figueiras and Y. Pazos, *J. Plankton Res.*, 1991, **13**, 589.
152. T. A. Thorsen and B. Dale, *Palaeogeogr. Palaeoclimatol. Palaeoecol.*, 1998, **143**, 159.
153. F. Marret, J. Scourse and W. Austin, *Holocene*, 2004, **4**, 689.
154. B. Dale, *J. Sea Res.*, 2009, **61**, 103.
155. E. Breton, V. Rousseau, J. Y. Parent, J. Ozer and C. Lancelot, *Limnol. Oceanogr.*, 2006, **51**, 1401.
156. X. Liu and W. Wang, *J. Geogr. Sci.*, 2004, **14**, 219.

Scientific Challenges and Policy Needs

MICHAEL N. MOORE,* RICHARD OWEN AND
MICHAEL H. DEPLEDGE

ABSTRACT

A key determinant of the quality and sustainability of the coastal marine environment is the dramatic growth of the human population, in particular along the global coastal zone, over the course of the last century. Burgeoning population growth, often as a result of reduced infant mortality and migration from rural communities, has created unprecedented social and economic demands for food resources, both in fisheries and aquaculture, while poor governance in respect of haphazard urbanisation and industrialisation and poorly regulated waste management have contributed extensively to the degradation of coastal ecosystems. Human health and wellbeing are consequently at risk from the resultant increased burdens of bacterial and viral pathogens from sewage and agricultural faecal run-off, as well as chemical and particulate waste from a variety of sources such as industry, domestic effluent, combustion processes, agricultural run-off of pesticides and nutrients, transport and road run-off. Unless policy formulation recognises that expansion of the human populations is often a key causative factor in the degradation of the coastal marine environment and related human health risks, and develops effective sustainability and mitigation strategies to deal with this, then any other actions will only provide expensive stop-gap solutions that are essentially 'papering over the cracks'. A recognition of the complex nature of the connectivity of the coastal marine environment with public health is critical for understanding the relationships involved.

A holistic systems approach such as Integrated Coastal Zone Management is necessary to address the highly interconnected scientific challenges of

---

*Corresponding author

increased human population pressure, pollution and over-exploitation of food (and other) resources as drivers of adverse ecological, social and economic impacts, and the urgent and critical requirement for effective public health solutions to be developed through the formulation of politically and environmentally meaningful policies. Since coastal zone environmental problems and related health and socio-economic issues are trans-national in character, the demands on regulation and governance go well beyond the actions of a single government and will require integrated action on a regional and global scale by national governments and stakeholders (*e.g.* non-governmental organisations), regional organisations (*e.g.* European Union) and international organisations (*e.g.* United Nations).

1 Introduction

Living organisms influence the environment in which they live and humans are no exception. In fact, human impact on our environment is, to an unusually large extent, a by-product, shaped by our social actions, governance, economic forces, international trade and several other factors.[1,2] In many cases, we are still not even aware of all those interconnections, and of how actions in one place affect other areas of our ecosystem. Therefore, environmental changes tend to be regarded as 'unavoidable' or as the 'unforeseen' consequences of profound economic and cultural changes that are being 'brought on to us' without 'anybody having the possibility or the ability to do anything about them'. Here we argue that there is much we can do through policy interventions, to manage the challenges we face resulting from humans interacting with the marine environment. The connectivity between human health and human ecology is far from straightforward, and factors that have a negative influence on ecosystem function and ecological integrity will not necessarily be directly linked to adverse effects on human health or wellbeing.[3,4]

To date, public health has mostly functioned with a simple univariate action–reaction model or, at most, multilevel relations but always with clear directionality.[2,4–6] For example, following identification of an algal toxin in seafood, action will be taken to ensure that consumption of the seafood ceases until the threat has passed. Interactions are much more complex in ecology and models of ecological and health interconnectivity will require a much greater degree of sophistication (see Figure 1).[7] This complexity underlines the need to adopt systems-level approaches when addressing the relationships between human ecology and public health issues (see Figure 2).

The marine environment is generally recognised as providing major environmental goods and services.[5,8] However, the pressures exerted on the coastal marine environment by expanding human populations are responsible in considerable part for the problems of habitat degradation, over-exploitation of biological resources, collapse of fisheries, and chemical and microbiological pollution.[6,8–14] All of these problems can combine to have a negative impact on public health in the areas affected (see Figure 1).[15]

Conceptual Model for Integrating Natural & Social Systems

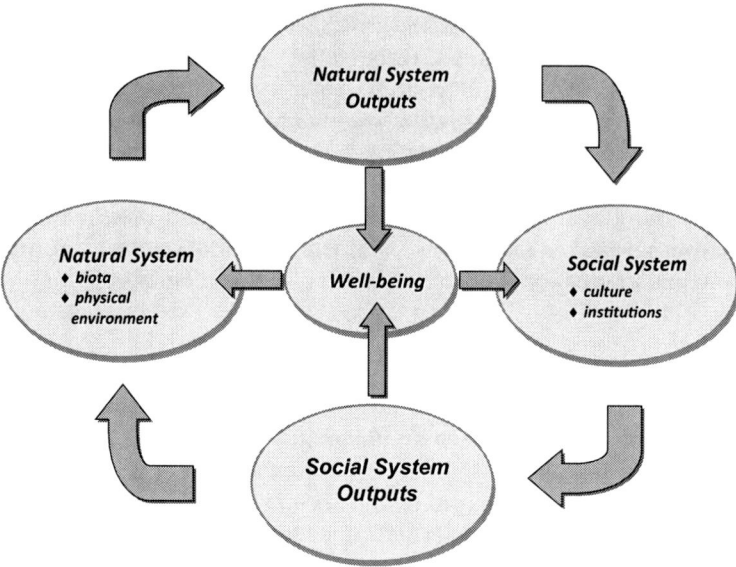

Figure 1 Generic conceptual model for integrating natural and social systems in the context of environmental health. (Adapted from Di Giulio and Benson).[3]

River basins, deltas and estuaries are often characterised by a rich diversity of plants and animals, or by particular assemblages of species that are, unfortunately, often environmentally sensitive and susceptible to human interference.[5,13,16] This can lead to conflict over resource rights and deprive the indigenous human population of major sources of food (*e.g.* fish and shellfish as a major source of protein). However, there is now increasing awareness of the global importance of specific geographical domains, such as the coastal land–sea interface, as major resources and concern for maintaining the diversity of life on our planet. This was a major focus for Agenda 21 of the UNCED, Earth Summit Conference in Rio de Janeiro in 1992.[17]

A significant proportion of the human population (up to 70%, depending on the definition of 'coastal zone') lives in close proximity to the coastal zone, often in the vicinity of large estuaries and river deltas, and is frequently dependent on the fishing industry for food, employment and wealth generation.[5,16,18,19] Consequently, the coastal and shelf areas are the most vulnerable zones of the ocean, since they are the most heavily used regions of the planet, often with fisheries and extensive urban and industrial development.[10,13,16,19] They receive a multitude of human and zoonotic pathogens, biogenic and chemical waste inputs originating from industrial, domestic and agricultural land-based sources which are difficult to estimate quantitatively.[20] A complex mix of other toxic chemical pollutants is also introduced through shipping

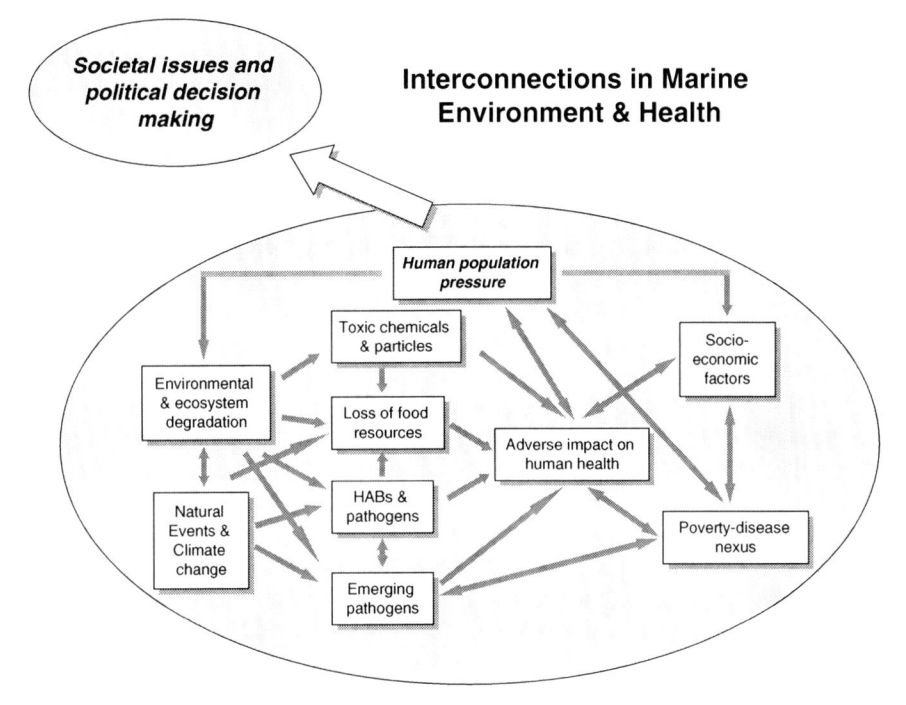

Figure 2 Hypothesised network of major processes contributing to the interconnectivity of the natural environment and human health that can be used to inform policy formulation. This is probably over-simplistic but serves to indicate the high level of connectivity from which the complexity of the system emerges. (HABs = harmful algal blooms). For additional detail see *Millennium Ecosystem Assessment*, 2005, http://www.millennium-assessment.org/en/index.aspx; and World Health Organisation, http://www.who.int/topics/environmental_health/en/.[132,133]

activities, offshore petrochemical industries and atmospheric inputs of airborne particles of industrial origin.[5] To further complicate this already complex situation, coastal zones include the most diverse and productive ecosystems in our oceans. Relatively recent economic studies have placed a higher value on coastal zones (US$12.6 trillion per year for coastal zones out of a global total of US$33.3 trillion per year) than any other compartment of our environment.[21]

Deterioration of the global coastal zone can be attributed to a variety of reasons, including: economic failure; inadequate governance and non-enforcement of existing environmental protection laws; haphazard industrialisation and urbanisation resulting in run-off of polluted wastewater and contamination of land, rivers and coastal waters; poor public education and understanding of the problems; and the strongly sectoral structure of government bodies frequently presents a barrier to integrated solutions.[8,9,18,22] Already high and/or increasing population density in the coastal zone often lies at the heart of these problems (see Figure 2).[23,24]

Pollutant impact on ecosystem function and human health is an urgent and international issue, since there is an ever-increasing number of examples of environmental disturbance, likely to affect the biota and humans, by both natural and anthropogenic stress.[5] Important stressors include domestic sewage and agricultural faecal waste run-off, toxic industrial chemical contaminants, medical and veterinary pharmaceuticals, radionuclides, natural biogenic toxins (*e.g.* from harmful algal and cyanobacterial blooms or HABs), sound and light pollution, climate change and increased UV radiation, nutrient enhancement or deprivation, hypoxia, habitat disturbance and pathogen-induced disease (see Figures 1 and 2).[6,13,14,16,25] In fact, environmental disturbance will frequently comprise various combinations of such stresses.[3] It is increasingly recognised that assessment of the impact of environmental disturbance on organisms, including humans, requires understanding of stress effects throughout the hierarchy of biological organisation, from the molecular and cellular to the organism and population levels, as well as the community and ecosystem level.[3,26] In the past, damage to the environment has largely been identified retrospectively and in response to acute events such as major disasters (*e.g.* industrial accidents like Seveso and Bhopal, oil spills such as the Amoco Cadiz, Exxon Valdez and Gulf War, and chemical pollution of the Great Lakes).[13,26] Generally, these have been measured in terms of human health impacts and visible changes resulting from the loss of particular populations or communities. However, long-term and chronic exposure to environmental stress, including chemical pollutants or other anthropogenic factors, will seldom result in rapid and catastrophic change.[5,26] Rather, the impact will be gradual, subtle and frequently difficult to disentangle from the process and effects of natural environmental change. This latter problem has been a major stumbling block in assessing environmental impact since such investigations began, mainly in the 1960s. In terms of tangible consequences, insidious environmental degradation may be an important contributing factor leading to reduced wellbeing and a poor quality of life, which are themselves associated with an increased risk of diseases.

There is undoubtedly a complex inter-relationship between human health and the oceans (see Figure 2).[5] Historically, research and regulatory concern have concentrated on the impact of human activities on the marine environment, particularly through anthropogenic pollution and the over-exploitation of resources. Considerable changes have already occurred in the ecology of many coastal regions as a result of both natural and man-made influences, largely linked to industry and urbanisation. In the future, further adverse changes are also likely to be brought about by sea level rise, ocean acidification and increased harmful UV radiation.[10,13,16,17,27] To further exacerbate the situation, continuing problems of water supply and sanitation, associated with pollution and water-related diseases, may contribute to continued human deprivation, sickness and death (see *Millennium Ecosystem Assessment*, 2005, http://www.millenniumassessment.org/en/index.aspx; World Health Organization, http://www.who.int/topics/environmental_health/en/; and Figure 2).[19]

Nonetheless, the situation is not entirely negative as the marine environment is also responsible for providing potential health benefits through the provision of food, novel pharmaceuticals and related products derived from marine organisms, and through general wellbeing from a close association with the coastal environment (*i.e.* recreational and psychological benefits – 'the Blue Gym' effect).[6,28,29]

The world's coastal regions, as mentioned above, are often characterised by a very rich diversity of plants and animals (biodiversity); particularly those situated in the tropics. In spite of their abundant natural resources, these areas are often environmentally sensitive and fragile regions owing to their particular natural physical setting and their distinctive ecological features and functions. Considerable changes are already occurring in the ecology of many coastal regions as a result of both natural and man-made changes, largely linked to increasing human population pressure and the associated habitat damage.[9,30] These include upstream dam construction, coastal zone modification and erosion, planned and haphazard urbanisation, forest clearance, agriculture, fisheries, industrial development, and oil and gas production. Such, often conflicting, pressures have resulted in the environmental degradation of many estuarine and delta regions. Unless this trend is reversed, it will undoubtedly lead to increased conflicts over resource rights and deprive indigenous populations of their main source of protein (*i.e.* fish and shellfish), leading to the prospect of malnutrition and starvation.

Consumer-driven global consumption of seafood and marine products is also placing increasing burdens on the resource capability.[8,9,12,16,31,32] New evidence linking fish-derived dietary products and health are also fuelling the demand for new retail products, such as fish oil and shark cartilage, many of which have either limited (*e.g.* omega-3 fatty acids in fish oil appear to have some benefits for cardiovascular disease) or no verified health benefits.[33,34] Ultimately, this pattern of behaviour and demand is unsustainable, although some resources may be increased through aquaculture in some specific instances.[31]

This chapter addresses the scientific challenges and regulatory policy needs facing the global human society in the context of our marine environment. This is self-evidently not a national or even a regional problem, but is in fact a major global issue that will require trans-national solutions if the marine environment is to remain ecologically functional and economically sustainable.[13] The consequences of past and current failure to stem the growth and expansion of the human population in ecologically vulnerable areas is evident on a global scale, and the resultant impacts on the terrestrial and marine environments has been catastrophic in many instances, with the loss of natural coastal defences, such as mangrove forests, and the coastal fisheries that supported the indigenous populations.[8,9,35–37] The associated health issues include potential starvation and famine through the loss of dietary resources, water-borne and seafood-borne pathogenic diseases, and chronic sub-lethal poisoning through chemical pollution (see Figure 2).[5,6,13,14]

In addressing these issues we need to consider what is at risk:

- Habitats and environmental resources (*e.g.* salt marshes, coral reefs, large estuaries and deltas, which are under pressure from unprecedented land use and urbanisation at a time of sea level rise and coastal erosion);

- Coastal biodiversity of land and aquatic animals and plants (*e.g.* loss of species from coastal rainforests, mangrove forests, fish and plankton populations are declining in some areas);
- Environmental quality (*e.g.* due to chemical contamination). Shellfish and fish-growth studies and rapid 'clinical-type' biomarker tests (*e.g.* Rapid Assessment of Marine Pollution, RAMP) can show where there are areas of contamination and harmful effect;[38]
- Ecotoxicology and environmental health (*e.g.* contaminants with endocrine-disrupting properties leading to sex changes in some animals and reproductive pathologies, carcinogenic and mutagenic compounds impacting on plants, animals and humans);
- Aesthetics (*e.g.* oil and litter on beaches);
- Loss of 'ecosystem services' (economic, cultural, *etc.*) as a result of these pressures and impacts;
- Effects on quality of life, including the economic, social and health implications of the deterioration of the marine environment.

In order to develop a strategy for dealing with the problems, we need to ask how can we forecast and reduce risks?[10,13,16] Research is needed into specific problems to find the most cost-effective solutions. In particular, we need:

- Innovative monitoring and surveillance techniques to understand the extent of the problems (*e.g.* earth observation systems for HABs, detection of chemical pollutants, biogenic algal and microbial toxins and human pathogens, improved testing for seafood safety);
- Research into understanding processes (*e.g.* physical, chemical and biological processes involved in the environmental transport and transmission of toxic chemicals and pathogenic organisms to humans);
- To develop environmental models to determine the extent of high natural dispersion areas around sewage and industrial discharge points;
- To develop expert systems to link existing models with our experience and knowledge of the environment;
- To develop and use indicators to show effectiveness in moving towards sustainable development where there is a need to link environmental, social and economic measures;
- To develop and implement methods that demonstrate the value (economic and otherwise) of marine environments at a local, regional and global scale;
- To develop an understanding of the direct and indirect causal relationships between degradation of the marine environment (especially in coastal regions) and the incidence of diseases and adverse effects on wellbeing of the human population.

By effectively identifying and interconnecting the interdisciplinary elements, we will see the emergence of new ways of solving problems in what at present are seemingly unrelated areas of environment and human health (see Figure 2).[3]

This chapter is not intended to be a comprehensive review of health concerns in the marine environment. Rather, it is aimed at identifying potential issues and raising debates on the issues that have received in-depth coverage by other chapters in this book and by other authors.[5,6,14,16,39]

2 Key Science Challenges for Marine Environment and Human Health

There is a pressing need to better understand how man-made and natural changes to the marine environment can influence human health through studying the complicated mix of environmental, social and economic factors that influence health.[3,4,40] These need to focus particularly on naturally occurring biogenic toxins, man-made pollutants, nanoparticles and pathogens to see:

- How they spread within the marine environment;
- How their properties change as they interact with other substances or organisms;
- How people become exposed to them, and their impact on human health;
- How they will affect human health and wellbeing in the future as the population undergoes demographic change (especially ageing), and under the more general influence of climate change.

2.1 Linking Ecosystem Integrity, Ecosystem Services and Human Health

Humans have altered, and will continue to alter, their environment, while remaining dependent upon ecosystems as resources of food, water and materials. Such alterations are a result of combinations of physical, biological and socio-economic factors.[2,3] However, until recently, evaluation and management of the resultant impacts on ecosystems, the services these provide and human health have historically been undertaken as largely separate activities, under the auspices of different disciplines with no obvious interaction. Hence, many of our perceptions of the relationships between the natural environment and human health are very limited and still relatively unexplored. These limitations have resulted in a knowledge gap for those seeking to develop effective policies for sustainable use of resources and environmental and human-health protection.[3,5]

In attempting to fill these gaps in our knowledge, our understanding of the functioning of the biosphere and our connections with it must be adequate in order to inform policy and decision-making processes. Unravelling the complicated network of interactions is a prerequisite if we are to develop a practical predictive capability to forecast how environmental change impacts on linkages between natural systems, social systems and human health (see Figures 1 and 2). The last 50 years have seen some progress in terms of understanding such interactions, but such predictive capability must remain a major longer-term scientific goal.[3,13,14,26] The scope of current capabilities will be, of necessity, more pragmatic and focus on capacity-building and exploratory research.

Undoubtedly, new developments and improvements in our scientific understanding of how environmental change impacts on the linkages between ecological integrity, environmental goods and services, and human health will aid us as we seek to develop an acceptable standard of living for many more people. This will in turn help us to ensure that the ecological pillars, which support our society and industries, are protected and remain sustainable.[3,8,9,30] We must also aim to successfully integrate social and natural systems on a local scale, while understanding the larger scale ramifications and consequences of decisions on a local, national and trans-national scale. Foremost in this context should be concern about limiting the adverse consequences of human population expansion (including migration from non-coastal areas) and how best to encourage limits to family size.[41] The family size issue will probably self-regulate with increasing economic improvement and personal wealth in the society.[19] However, it should be noted that although population growth is the major problem, there are coastal populations in some remote areas that are in decline and this too may lead to environmental problems.[42] For example, depopulation can lead to inefficient operation of sewage treatment facilities designed for a higher capacity, or to neglect of coastal defences and port facilities. Economic decline in areas suffering population decline carries with it a significant burden of disease, often reflected in big incidences of obesity and depression.

2.2 Sustainable Industrial Development

The major issues of concern include: the role of 'Industry' as a significant source of pollution; the fact that pollution does not respect national boundaries; the loss of living resources and biodiversity; damage to human health; and support for sustainable financing and banking in order to support developing economies (ICS-UNIDO, International Centre for Science and High Technology, United Nations Industrial Development Organisation; http://www.ics.trieste.it and http://www.UNIDO.org).[10] The environmental objectives of sustainable industrial development include the sound management of natural resources; effective transfer of environmentally sound technologies in order to reduce, re-use and recycle waste; investment promotion for sustainable industry; and environmental monitoring and control of investments for environmental industry projects.[10,13] The environmental components for sustainable industrial development need to include:

- An effective environmental policy framework;
- Cleaner industrial production and pollution prevention;
- Development and enforcement of environmental emission and discharge standards;
- Enforceable pollution control and waste management;
- Ecotoxicology for assessing environmental impact of pollution and overuse of resources;

- Environmental modelling for policy decisions;
- Effective risk governance – risk assessment and risk reduction;
- An effective integrating process with socio-economic conditions and governance issues.

Effective implementation requires knowledge-based expertise on environmental policy, cleaner production technologies, waste management and pollution control in order to achieve sustainability (ICS-UNIDO; http://ics.trieste.it and http://UNIDO.org).[10,13] It also requires expert diagnostic and predictive software to link existing models with physical and chemical information, and knowledge of the environment. Additional requirements include the development and use of indicators, which show effectiveness in moving towards sustainable development, that link environmental, social and economic measures.

2.3 Understanding and Mitigating the Impacts of Climate Change

Weather and climate factors are known to affect the transmission of water- and vector-borne infectious diseases, as well as the transport of chemicals around the environment.[5,43] Climate change may therefore have important impacts on the dispersion of pathogens and chemicals in the environment. In addition, changes in climate are likely to affect the types of pathogens occurring, as well as the amount, type and physico-chemical speciation of chemicals entering the marine environment for a range of scenarios.[44-48] Future risks from pathogens and chemicals could therefore be very different than they are currently. As a consequence, is important that we begin to assess the implications of climate change for changes in human exposures to pathogens and chemicals, and the subsequent health impacts in the near and longer-term future.[6,43] Coupled to this are indirect health threats, for example, from climate-related coastal flooding that may threaten health directly, but also indirectly as a longer-term insidious threat for many months, or even years afterwards, in the form of psychiatric disorders.[49]

Climate change is likely to increase human exposure to agricultural contaminants.[43] The magnitude of these increases will be highly dependent on the contaminant type. Risks from many pathogens, particulate and particle-associated contaminants could increase significantly. However, these increases in exposure could, for the most part, be managed through targeted research and policy changes.

2.4 Better Prediction Systems for Natural Disasters

Natural catastrophes caused by earthquakes, volcanic eruptions, cyclones and hurricanes, and tsunamis are recurring events with variable predictability. While attracting disproportionate media attention they can still cause significant ecological damage and impact severely on human health in affected locations (UNISDR, United Nations International Strategy for Disaster Reduction; http://www.unisdr.org).[19] However, the adverse ecological effects often tend to be short-term and recovery is frequently quite rapid. Apart from the loss of lives and livelihoods, the impact on human health is often temporary in broad environmental terms. This is to no way underestimate the undoubted human misery caused by such events, however, and the longer-term psychological and psychiatric problems that may arise.[49]

Some indirect health impacts may arise from natural disasters, nevertheless, through the release of toxic waste and human pathogens arising from the inundation of coastal landfill sites and agricultural land, and the disturbance of estuarine sediments.[19,48,50]

Improvements in modelling capabilities coupled with remote-sensing should help to provide better early-warning systems in the future.

2.5 Understanding the Distribution and Risks of Marine Biogenic Toxins (Algal Toxins)

Global consumption of marine products continues to expand, particularly in Asia.[5,6,39,51,52] With the ensuing global decline in wild fish stocks, it is inevitable that the aquaculture industry will continue to grow. Consequently, it is critical to ensure the safety of seafood products, both farmed and wild, without undermining public confidence in it.[6,14,25,39,51–53]

Monitoring for biogenic algal toxins in the British, European and North American waters has been generally successful in safeguarding humans from shellfish poisoning. However, while the threat posed by cyanobacterial toxins from inland waters has been recognised by medical practitioners, the risk associated with marine biotoxins is less well appreciated.[25,52] Although this may have led to an under-reporting and recording of shellfish poisoning events, it may equally have contributed to an unjustified but common public perception of shellfish consumption as being 'risky'. Quantitative evidence of human intoxication levels and an understanding of the magnitude of any health risk set against the health benefits of shellfish consumption is therefore a challenging but important research priority. Furthermore, while an increasing ability to detect and quantify toxins is reassuring, it is important that research to quantify the concentrations of these toxins at which they are medically significant to humans keeps pace, preventing 'scaremongering' or unnecessary harvesting closures.[6,52]

As harmful blooms typically develop rapidly, risk assessment methodologies are required to allow the industry to better plan harvesting operations at times of lower risk.[23,52] While it is clear that some general patterns exist, harmful phytoplankton exhibit spatial and temporal variability. Such heterogeneity suggests that local risk assessments based on detailed knowledge of the physiology of the causative species, and hydrography and meteorology of the local environment are most likely to be successful.

Finally, there remains a need to be vigilant for invasive species. Given the global increase in sea surface temperatures, as well as the potential for the introduction of new species *via* ship ballast, it will be important for monitoring agencies to familiarise themselves with the identification of potential invasive species.[18,52]

2.6 Identifying and Reducing Viral and Bacterial Pathogens from Sewage and Agricultural Run-Off

Many human pathogens have reservoirs in the environment or are transmitted between humans by animal vectors or through animal intermediate hosts.[6,14,43] Better control of human pathogens requires an understanding of their transport and ecology in the environment. It is also important to try to anticipate new emerging diseases, a problem that is likely to become acute with global climate change and increasing globalisation, with its concomitant rapid transfer of people and products throughout the world.

Transfer of these pathogens *via* faecal waste from farmed animals into waterways and hence into the marine environment is a potentially important hazard, particularly with the current trends for intensification of livestock farming, which can result in vastly greater amounts of waste entering the aquatic environment.[24,43,55]

Furthermore, irrigation of crops with contaminated water or organic waste is a potential means of contaminating foodstuffs and waterways with enteric viruses, and studies have demonstrated that viruses can be transferred to the surfaces of vegetables and persist there for several days, following the application of sewage sludge or effluent.[24,56,57] These studies were all performed outside the UK, and survival of viruses in the environment or on crops under UK climatic conditions has not been fully determined experimentally.[57]

2.7 Understanding Emerging Risks (e.g. Nanoparticulates from Industrial and Domestic Use)

Nanotechnology is a major innovative scientific and economic growth area which may present a variety of hazards for environmental and human health.[58,59] The surface properties and very small size of nanoparticles (NPs) and nanotubes provides surfaces that may bind and transport toxic chemical pollutants, as well as possibly being toxic in their own right by generating reactive radicals.[60,61] There is a wealth of evidence for the harmful effects of nanoscale, combustion-derived particulates (ultrafines) which, when inhaled, can cause a number of pulmonary pathologies in mammals and humans. However, release of manufactured nanoparticles into the aquatic environment is largely an unknown. Possible nanoparticle association with naturally occurring colloids and particles needs to be considered, together with how this could affect their bioavailability and uptake into cells and organisms.[58] Uptake by endocytotic routes have been identified as probable major mechanisms of entry into cells, potentially leading to various types of toxic cell injury.[62–65] The higher level consequences for damage to animal health, ecological risk and possible food chain risks for humans should be considered as well, based on known behaviours and toxicities for inhaled and ingested nanoparticles in the terrestrial environment.[66–68] A precautionary approach is probably advisable, with individual evaluation of new nanomaterials for risk to the health of the environment. Although current toxicity testing protocols should be generally applicable to identify harmful effects associated with nanoparticles, research into new methods is required to address the special properties of nanomaterials.

It is likely that the same factors responsible for the novel properties of nanoparticles may be the source of their potential hazard.[66–68] Discrepancies between existing toxicological studies have shown the importance of good quality, well-characterised nanomaterials for toxicological studies, combined with reliable synthesis protocols.

It is important to obtain complete characterisation, including the recognition of impurities and surface properties.[66–68] For regulatory purposes, a precautionary approach has been recommended by the Royal Society and Royal Academy of Engineering.[69] This will probably require that each type of new nanomaterial should be treated individually for toxicity and risks to the health of the environment, as it is not feasible to generalise about the toxicity of nanoparticles. Even though existing nanomaterials are very diverse in their composition and surface properties, current toxicity-testing protocols should still be generally applicable to identify harmful effects associated with nanoparticles.[70] Proposed new regulatory frameworks for chemical risk assessment procedures such as the European Union's REACH (Registration, Evaluation and Authorisation of Chemicals) may be suitable for adaptation to include nanomaterials (http://europa.eu.int/comm/environment/chemicals/reach.htm).

Since there is so little data available for aquatic environments, research is required to test the behaviour and particulate-binding properties of manufactured nanoparticles in both freshwater and seawater (salinity may alter the surface characteristics of nanomaterials). The relative importance of the endocytotic and caveolar routes of uptake identified above also needs to be assessed in representative aquatic species, since this will be a crucial factor governing intracellular behaviour, distribution, fate and toxicity of internalised nanomaterials.[71]

Accumulating evidence on the toxicity of nanoparticles (NPs) indicates that free-radical-mediated oxidative stress is a generic reaction, although, as yet, the impact of the products of nanotechnology on animal and ecosystem health is largely an environmental unknown.[66–68] Animal cells are evolutionarily pre-adapted to internalise nanomaterials *via* invagination at the cell membrane into cytoplasmic vesicles (cell feeding or endocytosis).[58,62–65] Hence, most uptake of NPs will probably occur *via* this route. Nevertheless, fundamental questions remain unanswered as to the significance of NPs on the health of aquatic organisms; the interactions between NPs and conventional chemical pollutants, such as organic xenobiotics (*e.g.* PAHs and heterocyclics); the higher-level consequences for damage to animal health; the associated ecological risk; and the possible food-chain risks for humans.[58,62,72]

The toxicological and pathological interactions of complex mixtures of environmental contaminants and natural particles, such as can occur in the real environment, are still a largely unknown quantity. This gap in our knowledge needs to be addressed in order to extend the database for developing risk assessment procedures for nanomaterials in the aquatic environment.

A major challenge for environmental toxicologists will be the derivation of toxicity thresholds for nanomaterials, and the determination as to whether currently available biomarkers of harmful effect in aquatic ecosystems and humans will also be effective for environmental nanotoxicity and nanopathology.[73] If new methods are required to assess the toxicity of nanomaterials, then these tests will also need to be linked, if possible, with functional ecosystem indices.[74–77] Such linkage would be desirable in order to bridge the gap between individual organism 'health-status' and ecosystem-level functional properties, and how they affect and are quantifiably connected to the health of the environment.[58,67,68,74–76,78]

2.8 Conventional Chemical Inputs (Industrial, Domestic, Agricultural and Road Run-Off), including Personal Care Products, Disinfectants, Pharmaceuticals, Novel Chemicals and Radionuclides

Environmental impacts by both natural events and man-made interventions are a fact of life, and developing the capacity to minimise these impacts and their harmful consequences for biological resources, ecosystems and human health is a daunting task for environmental legislators and regulators. Man-made chemical inputs include industrial, domestic, agricultural nutrients and pesticides, road run-off, personal care products, disinfectants, pharmaceuticals and novel chemicals; these have been comprehensively covered in Chapter 3 and elsewhere.[56,79]

A major challenge in impact and risk assessment, as part of environmental management, is to link harmful effects of pollution (including toxic chemicals) in individual sentinel animals to their ecological consequences. This obstacle has resulted in a knowledge gap for those seeking to develop effective policies for sustainable use of resources and environmental protection. Part of the solution to this problem may lie with the use of diagnostic clinical-type, laboratory-based ecotoxicological tests or biomarkers (*e.g.* Rapid Assessment of Marine Pollution, RAMP), utilising sentinel animals as integrators of pollution, coupled with direct immunochemical tests for contaminants.[11,38] These rapid and cost-effective ecotoxicological tools can provide information on the health-status of individuals and populations based on relatively small samples sizes. In the context of health of the environment, biomarkers are also being used to link processes of molecular and cellular damage through to higher levels (*i.e.* prognostic capability), where they can result in pathology with reduced physiological performance and reproductive success.[80] Although some studies exist,[11,81–84] there is still a significant but scientifically challenging lack of understanding for interactions between various environmental stressors and their ecological consequences.

Complex issues are involved in evaluating environmental risk, such as the effects of the physico-chemical environment on the speciation and uptake of pollutant chemicals and inherent inter-individual and inter-species differences in vulnerability to toxicity, and the toxicity of complex mixtures.[44,46,85–87,89,90] Effectively linking the impact of pollutants, through the various hierarchical levels of biological organisation to ecosystem and human health, requires a pragmatic integrated approach based on existing information that either links or correlates processes of pollutant uptake, detoxication and pathology with each other and higher level effects.

Generic quantitative models may provide a partial solution to the problem of predicting the dynamics and risk of adverse impacts of chemical pollutant stressors, through the development and implementation of computational models.[91,92] Such models could be used to simulate pollutant pathways in the human food chain, and could be adapted to environmental transport of human pathogens and their viability, and the interactions between pollutants and microbial populations that may lead to increased biogenic toxin production. The kinetics and dynamics of harmful carcinogenic and toxic chemicals in fish and shellfish species that are harvested for human consumption could also be characterised and modelled, as could pathological reactions (molecular to tissue level) to known pollutants and potentially hazardous novel particles, chemicals and radionuclides.[93–96] The systems approach is considered to be essential if we are to effectively interpret reductionist chemical and biological data in a meaningful holistic context, for prediction of risk to consumers of seafood and recreational users of the marine environment.[40] This step is necessary to ensure that the social and economic implications of reductionist chemical and biological data are understood, and to encourage the application of these data in an environmental management context.

This approach will allow us to explore the use of measures of environmentally induced pathological deterioration in flesh quality or 'health status' of seafood, particularly filter-feeders (bivalves) and bottom-dwelling fish, due to pollutants or pathogens as predictors of risk to human health.[11,26,97–99] A methodological co-evolutionary 'synthesis through modelling' approach will facilitate targeted experimental design and field sampling, as well as the effective integration of multiple environmental datasets and their subsequent interpretation. This synthesis is viewed as a crucial step towards the derivation of explanatory frameworks for prediction of outbreaks of human pathogen-related diseases, potential for low-level chronic exposure to biogenic and anthropogenic chemical contaminants, as well as environmental impact on animal health status as a surrogate measure for human health risk.[6,11,14,96,98,100] Therefore, the novel use of biogeochemical and molecular measurements of pollutant kinetics and dynamics, high-throughput molecular methods for tracking pathogens and determining their viability, and new probes for molecular, cellular and tissue pathological reactions to environmental stressors, coupled with simulation modelling, is proposed as a practical approach to the development of a pre-operational toolbox.[11,26,82,92,98,101] The overall goal should be to provide a pre-operational toolbox that will facilitate an environmental management regime that maintains both food security and appropriate public health management for human use of the marine environment and aquaculture as a resource for human food, not only for the UK but internationally.[14,102]

Emerging pollutant issues, such as changes in environmental burdens and effects of contaminants with climate change, and substances with 'new' mechanisms of action (endocrine disruption, reproductive and developmental toxicity), ensure that chemical contaminants remain high on the agenda of environmental threats. A range of readily applicable techniques is now available for assessing pollutant impact, but these need to be used to manage marine resources. In the *EU Water Framework Directive*,[103] there is an emphasis on community and biodiversity as measuring sticks for human impacts in freshwater and marine ecosystems.[26,82,102,104,105] However, it is unlikely that these tools alone will provide the necessary information required for effective environmental management. Consequently, it would probably be more effective if ecological indicators such as biodiversity were used alongside and integrated with rapid assessment methods (*e.g.* biomarkers) that are more sensitive to early changes and more specific to impacts from human activities. The current methods for environmental risk assessment tend not to include data for adverse health effects in the biota, such as is provided by some biomarker approaches (*e.g.* RAMP) described above.[11,81,98]

2.9 Endocrine Disruption

Anthropogenic chemicals which can disrupt the hormonal (endocrine) systems of wildlife species are currently a major cause for concern.[101,106,107] Reproductive hormone-receptor systems appear to be especially vulnerable. In the past few years, numerous effects of endocrine-disrupting chemicals on wildlife have emerged, including changes in the sex of riverine fish, reproductive failure in birds and abnormalities in the reproductive organs of alligators and polar bears. Much less is known regarding endocrine disruption in marine invertebrates, the key structural and functional components of marine ecosystems.

Potential effects of different classes of endocrine-disrupting chemicals in aquatic invertebrates have been reviewed by Depledge and Billinghurst.[106] Examples of endocrine disruption in marine invertebrates include imposex in gastropod molluscs exposed to organotin compounds and intersex in crustaceans exposed to sewage discharges.[106]

These authors considered the comparative endocrinology of several major invertebrate groups in order to try to identify which phyla were most likely to be at risk.[106] They also pinpointed gaps in our knowledge concerning the availability and uptake of endocrine disruptors. For example, the relative importance of different routes of chemical uptake (from seawater *via* food) was considered to be a key factor; including the feeding strategies (herbivores, carnivores, deposit feeders, suspension feeders) of various groups of animal.

2.10 Pharmaceuticals from the Sea

Some marine organisms are known to produce secondary metabolites with pharmaceutical potential.[6] Globally, the marine pharmaceutical pipeline consists of three US FDA-approved drugs, one EU-registered drug and 13 natural products at various stages of clinical testing.[108]

The technical and economical potential of using the marine biota for its therapeutic benefits is probably considerable but will require inputs from many areas of clinical, pharmaceutical and marine science if this potential is to be exploited in the future.

2.11 The Marine Environment as a Health and Wellbeing Resource: the 'Blue Gym' Effect

Marine environment and health have predominately been framed from the point of view of risks (of chemicals, pathogens and other stressors) to health. Literature in the field is dominated by studies identifying hazards and assessing and managing risks. Marine products that promote health (including pharmaceuticals, as mentioned in the previous section) provide an alternative avenue for research, but this is usually considered as a separate field of study. Far less research has historically been undertaken into the potential for the marine environment itself to directly promote health and wellbeing. This potential was realised at least as far back as the Victorian times when there was a fashion for patients to spend time in coastal sanitoria for certain conditions, but is now experiencing a resurgence through concepts such as the 'Blue Gym'.[28] The Blue Gym (http://www.bluegym.org.uk) is an innovative project originating in Cornwall (UK) but now spreading worldwide, utilising the coastal environment as a resource to promote health and wellbeing by increasing physical activity, reducing stress and building stronger communities. The hypothesis is that this may in turn encourage people to value and conserve the marine environment. The Blue Gym effect may demonstrate the importance of both a healthy natural environment and our own health by getting more people to be active outdoors using the sea, rivers, lakes and canals. There is evidence to show that people living near the coast are generally more active and that those who are not can be encouraged to become so.[28] We now know that contact with nature has additional mental health benefits that make it so important to maintain a clean and healthy water environment.[109] Similar health benefits are now well documented for people regularly using the green environment for walking, recreation and other activities. Interestingly, aquatic environments in general and coastal environments in particular seem to be strongly preferred to any others when evaluated in psychological assessments.[29]

3 Public Health Needs

3.1 Health-Related Indices of Environmental Impact

Indices of environmental impact should also be considered as a potential early indicator of harmful interactions with human health. A broad approach to the complex problem of assessing the 'health of ecosystems' will facilitate the validation and further the essential new development of robust and rapid tools for assessment that will provide indicators for public health risk.[11,26,74,98,102,110–112] Future efforts should focus on an integrated approach to the validation of biomarkers that are prognostic for ecological endpoints and that also have implications for human health, such as food safety.[11,14,26,74,80,98] As with bioavailability and uptake, exposure to pollutant mixtures must also be considered with the possibility of complex synergistic interactions resulting in emergent and novel toxicities and pathologies.[47,85]

3.2 Seafood Safety

Current regulatory frameworks for seafood safety lack sufficient integration for effective reduction of risk and do not account for the international nature of seafood trading and the nature of seafood production.[33,54,113–115] There needs to be a considerable improvement in the understanding of the public health risks associated with both farmed and naturally harvested seafood.[14,114]

The concept of a "chain of custody" – from the ocean to the final consumer – has been proposed by Yasusda and Bowen (2006) as a useful integrating framework for understanding and refining efforts to reduce public health concerns surrounding the consumption of seafood.[54,113]

In order to achieve this objective we need to develop a capability for predicting risks to human health by developing and adapting generic conceptual and computational models of pollutant and pathogen kinetics and dynamics, the influence of pollutants on microbial toxin production, and the pathological effects of carcinogenic/toxic contaminants on shellfish quality as a surrogate predictor of human health risk. These could then be used to test predictive parameters of progression of the decline in potential food quality and consequent increased risks to human health.

3.3 Environmental, Social and Economic Interactions (Quality of Governance, Overpopulation and Sustaining Critical Coastal Ecosystems)

Managing and mitigating the effects of human activities on the structure and function of coastal and freshwater ecosystems in a whole ecosystem context is emerging as a unifying theme for environmental protection, resource management and integrated environmental management.[19]

The overall problem is, how can we develop effective procedures for environmental/ecological impact assessment and risk evaluation, which also include human health, economic and other societal issues? One of the major difficulties in impact and risk assessment is to link harmful effects of chemical pollutants in individual sentinel organisms with the integrity of ecosystem function and possible resultant risks for human health.[5,6,14] Consequently, this obstacle has contributed to a 'knowledge-gap' for those seeking to develop effective policies for sustainable use of resources, industrial development and environmental protection.

There is a broad consensus that, as a society in Europe, we must change the ethos of how we do business. Domestic and industrial practices that were acceptable in the past will not be acceptable in the future, in large part because they will not be sustainable (ICS-UNIDO, http://www.ics.trieste.it and http://www.UNIDO.org). Changes will be required, and these changes will occasionally be dramatic and at other times will be slow and less overt. Changes will occur in the choices that we make as individuals, in the technologies and solutions to problems that we demand from industry and government, and in the role that communities, in the broad sense, play in decision-making.

3.4 Modelling – Need for an Integrated Approach in the Development of Effective Environmental and Public Health Policies on a Regional and Global Scale

In the past decade, genomic and proteomic technologies have spectacularly expanded the methodology and explanatory capability of the biological sciences. One of the next major challenges is to provide a mechanistic explanatory and predictive framework for the integrated expression of emergent function in whole systems such as cells, organs, animals and even ecosystems. System function emerges through the interactions among the components of the system in question (*e.g.* cellular proteins, cells, tissues, individual organisms, populations and assemblages or communities) and the integrative emergent properties or 'biocomplexity' of the system are 'computed' by these interactions. Consequently, the system can be viewed as a biological computer.[116–118] Therefore, if we want to explain complex physiological or ecological functions, as well as various pathologies and diseases induced by environmental stressors (including pathogens and pollutants), then we must follow nature's computational example and develop a predictive simulative capacity that exhibits emergent behaviours, alongside our existing experimental methodology.[78,88]

This new methodology of biological systems modelling for human health is heavily reliant on cell-based physiological and pathological simulation. These types of simulation provide an essential complement to experimentation and will increasingly play a central role in the investigation of normal function, responses to environmental signals (eco-physiomics), environmental toxicology and disease processes.[118–120]

Novel modelling tools are needed that will allow physiologists, epidemiologists, epizootiologists and environmental toxicologists to search for possible pathological targets for environmental stressors in a rapid and rational manner.[26,78,88,92,120] This search will be carried out in models of physiologically normal systems, as well as disease models, in order to facilitate prediction of relevant environmentally-related endpoints and synergies.

4 Policy Needs

The last 50 years has witnessed the steady growth of risk-based regulation as a key policy strategy for protecting the marine environment and human health *via* its implementation. Arguably it was Rachel Carson's book, *Silent Spring*, in 1963 that propelled appreciation of the widespread impacts of chemical pollution into the public consciousness, catalysing President Richard Nixon to establish the US Environmental Protection Agency nearly a decade later. Since the 1960s and 1970s, an increasing amount of legislation has been developed in the environment and health arena. Some of this has been specifically designed for protection of the marine environment (*e.g.* OSPAR for the North Sea) and various Directives applied across the EU for particular stressors (*e.g.* the *Bathing Water Directive* for water quality in the context of pathogens). Other pieces of legislation are not particularly focussed on the marine environment but are still very relevant (*e.g.* the REACH legislation and the Stockholm Convention on Persistent Organic Pollutants, UNEP – POPs). Some POPs, notably the halogenated organic compounds and some heavy metals such as mercury, have been found to travel great distances from sources of manufacture and application to high latitudes *via* atmospheric transport and to accumulate in polar marine food chains, leading to extensive contamination of wildlife and indigenous populations such as the Inuit. The latter requires hazard and risk information for industrial chemicals prior to their authorisation for use in the EU and includes environmental considerations.

REACH illustrates one of two features of policy and regulatory development over the last few decades. Regulation has taken an increasingly precautionary approach and, indeed, in the EU, policy is underpinned by an explicit commitment to the Precautionary Principle (European Community, Lisbon Treaty, Art. 191) and enacted through mechanisms such as 'data before market'; legislation enshrined in REACH and other Directives (*e.g.* those covering pharmaceutical products). The precautionary principle is both defined and applied in different ways.[121] A commonly cited version is that within the UN Rio 1992 Declaration: "where there are threats of serious or irreversible damage, lack of full scientific certainty shall not be used as a reason for postponing cost-effective measures to prevent environmental degradation."

The precautionary principle has been broadened under the ecosystem approach of the OSPAR Commission (http://www.ospar.org) to encompass all human activities: "By virtue of the precautionary principle, preventive measures are to be taken when there are reasonable grounds for concern that human activities may bring about hazards to human health, harm living resources and marine ecosystems, damage amenities or interfere with other legitimate uses of the sea, even when there is no conclusive evidence of a causal relationship. A lack of full scientific evidence must not postpone action to protect the marine environment. The principle anticipates that delaying action would in the longer term prove more costly to society and nature and would compromise the needs of future generations."

The second feature is best illustrated by perhaps the most important piece of environmental regulation covering the aquatic and near-shore environment to emerge in recent decades, the *EU Water Framework Directive* (WFD) (2000/60/EC).[103]

The WFD is underpinned by an integrated risk assessment and management approach, whereby there is consideration of both chemical and ecological status in defining water quality, with 'ecological status' incorporating biological, physico-chemical and hydro-morphological elements.[122] Water bodies (including rivers, lakes, estuaries, coastal waters and groundwaters) are periodically assessed and assigned to a classification system that grades their deviation (high, good, moderate, poor and bad) from a comparable site having no or very minor disturbance from human activity.[123] Programmes of measures (*e.g.* risk management) are put in place to facilitate meeting the WFD's overarching environmental objective for all water bodies to achieve 'good ecological' status by 2015. This illustrates a move towards more holistic, integrated risk assessment and risk management approaches, in which ecological goals and assessment methods play a more prominent and even central role. This integrated approach recognises the complex multifactorial nature of environmental risks, and the need to apportion these to specific pressures such as diffuse agricultural pollution and point source discharges from industrial complexes or sewage treatment works.

Within the coastal zone this integrated approach has been expanded further through the concept of Integrated Coastal Zone Management (ICZM), supporting regional transboundary-level programmes for inter-governmental bodies (EU – European Commission Coastal Policy, ICZM, http//ec.europa.eu/environment/iczm/pdf/evaluation_iczm_report.pdf).[13,16,110] This essentially 'systems approach' assesses the changing states of coastal ecosystems using science-based information, linked to socio-economic benefits for countries sharing or bordering on large estuaries and deltas. The methods can be used in an integrated interdisciplinary way in order to address the consequences of ecosystem change and the ensuing implications for sustainable use and development of food resources, as well as the needs of industry and the impact on human health.

ICZM holistically assesses the changing states of coastal ecosystems based on information obtained from five operational modules:

1. Ecosystem productivity;
2. Fish and fisheries (sustainability and seafood safety);
3. Pollution and health (both ecosystem and human);
4. Socio-economic conditions (poverty alleviation and public health improvement through development, education and investment);
5. Governance protocols.

These modules link science-based information to socio-economic benefits for countries sharing coastlines or bordering on large estuaries and deltas. The methods can be used in an integrated interdisciplinary way in order to address the consequences of ecosystem change and the ensuing implications for sustainable exploitation and development of food resources, as well as the needs of industry and the impact on human health. It is essential in the use of these methods to always strive to address the needs and welfare issues of regional populations, as well as the requirements of industry, fisheries and agriculture for sustainable economic development (EU – European Commission Coastal Policy, ICZM, http//ec.europa.eu/environment/iczm/pdf/evaluation_iczm_report.pdf).[10,37]

The methodology of 'integrated coastal zone management' brings together, by necessity, elements for dealing with the complex interactions of the many and varied demands placed on the coastal zone. The methods can be implemented through training (sharing of expert knowledge and technology transfers) and capacity building. These methods are firmly grounded on strategic science-based assessments and monitoring, linked to standard internationally agreed quality assurance protocols.

The development of such approaches and tools to support management decisions concerning the marine environment has been an important policy priority over the last 50 years, and will continue to be important in the future (*e.g.* current areas of interest include the development and use of health impact assessment, health damage valuation, human biomonitoring and cost–benefits methodologies as supporting tools).

Further policy priorities in the EU in the general area of environment and health policy have been articulated within the EU Action Plan on Environment and Health (CEC, Congress on Evolutionary Computation, 2004), which covered the period between 2000 and 2010. The marine environment is not treated as a special consideration within the Action Plan but we can identify key priorities (see Box 1), as these relate to the marine environments that are covered by it. These priorities include better understanding of the links between environmental quality and human health (particularly for vulnerable groups such as children); the development of tools to better characterise causal links (*e.g.* environmental health indicators, biomonitoring approaches); and the need for integrated monitoring approaches for the environment (including food) to allow the determination of relevant human exposure. Endocrine disruption is a specific area where more information is needed, as are the effects of chemical mixtures in the environment.[85,86]

Most recently a survey conducted by one of the authors as part of the EU Environment and Health ERANET (a European Research Area Network of environmental agencies, research funders and government departments across

Box 1 Some current environment and health policy priorities across the EU.

Common policy priorities in the environment and health area across the EU.
* = relevant to marine environment.
(Source: EnvHealth ERANET, http://www.era-envhealth.eu).

1. *Climate change: health impacts, health consequences
2. *Environment and health risks of (emerging) technologies: nanotechnology, environmental technologies, electromagnetic fields
3. *Tools and techniques: health impact assessment, biomonitoring, socioeconomic approaches, cost–benefit (*e.g.* health damage evaluation)
4. Indoor air pollution: cardiovascular health, aeroallergens, particulates
5. *Vulnerable or susceptible groups, environmental health inequalities: children, prenatal, social unevenness, environmental justice
6. *Endocrine disruption

Europe) confirmed that these remain priorities at an EU level, in addition to other priorities such as a better understanding of the risks of emerging technologies (*e.g.* nanotechnology); improved understanding of the health impacts and consequences of climate change; and risk management of priority (toxic) substances such as mercury, PAHs, benzene, chlorinated solvents and CMR substances (Carcinogenic, Mutagenic and Reprotoxic chemicals) (see Box 1).[82,87,112]

These policy priorities are very complementary and in some cases similar to the scientific challenges identified above, although from a policy perspective priorities remain predominantly focused within a risk assessment and risk management frame, with little consideration of some issues such as the marine environment as a health-promoting resource.

In March 2010 the 5th Ministerial Conference on Environment and Health convened under the auspices of the World Health Organisation in Parma, Italy, identified the 'key environment and health challenges of our time' as being:

- Health and environmental impacts of climate change;
- Health risks to children and other vulnerable groups posed by poor environmental, working and living conditions (especially the lack of water and sanitation);
- Socio-economic and gender inequalities in the human environment and health;
- Burden of non-communicable diseases, in particular to the extent that it can be reduced through adequate policies in areas such as urban development, transport, food safety and nutrition, and living and working environments;
- Persistent, endocrine-disrupting and bioaccumulating harmful chemicals and (nano)particles; and by novel and emerging chemicals and materials.

5 Discussion

Environmental stressors and their associated risks have always been an inherent part of society.[3] Ecosystems that encompass all modern human activity have been profoundly modified or completely altered, to both the detriment and the benefit of human health and society. Therefore, society must weigh the benefits against the risks associated with development and progress and make the sometimes difficult but necessary choices (International Risk Governance Council, 2010, http://irgc.org/IMG/pdf/irge_ER_final_07jan_web.pdf). However, the negative effects have not always been immediately obvious, given the difficulties inherent in 'future-proofing'. In many cases, the direct and indirect stresses of human and ecological health associated with urbanisation, industrial processes and pest control have been subtle, and we have been slow to recognise these stresses.[3] Since we did not understand human–ecological interactions, we could not calculate appropriate risk–benefit ratios!

Whereas considerable progress has been made to address the more obvious problems, there is concern that subtle effects of modern society, such as urbanisation, poorly regulated industrial development and habitat destruction, on human and ecological health may go undetected until more sensitive sub-populations or ecological indicators are adversely affected. Thus, in this chapter, we argue the case for the identification, quantification and application of biological indicators of ecological integrity, as well as the determination of the similarities and differences in responses of various aquatic ecosystems to stress.[77] Our objectives in this respect are two-fold:

1. We need to identify appropriate diagnostic and prognostic markers of health of the environment that might be used as sensitive indicators of the stresses that result in biological or ecological damage;
2. We need to reconcile various prognostic marker responses with ecological consequences that have a human health relevance.

This will require a systems-based approach, as already stated, involving the development of a hierarchy of interacting and overlapping computational models, in order to integrate those essential elements selected from the inherent complexity of environmental interactions that are necessary for effectively predicting ecological health risks.[78] Achievement of this goal will require the integration of functional ecological expertise with that of environmental physics, chemistry, toxicology and epidemiology, in order to provide insights into the relationship and prediction of the ecological and human health and possible societal risks associated with pollution arising from modern technologies.[26,124] As a sub-objective, it will also be necessary to derive new strategies for improving risk evaluation and try to avoid past problems, as well as countering the new pollution problems that will arise though the introduction of biotechnology- and nanotechnology-based industries.[26,58,88]

While it is recognised that healthy ecosystems provide basic goods and services to humans and other organisms, some of these are so basic (for example, air and water quality) that they are taken for granted when ecosystems are healthy.[5] However, when the buffering capacity of aquatic ecosystems is exceeded, as in many highly industrialised areas, there is frequent evidence of compromised human health, particularly in the developing world (*e.g.* Bangladesh arsenic problem and water-borne diseases).[10,13,14] It is probably a reasonable assumption that such phenomena have impacted other organisms as well.[5,11]

Unfortunately, the necessary epidemiological data for pollutant impact on sentinel animals, that should permit a comprehensive understanding of possible causal links between ecosystem integrity and human health, are often limited or fragmentary, both spatially and temporally.[4,14] Consequently, alternative interdisciplinary approaches will be required to identify such links, but interdisciplinary thinking must indeed extend beyond the boundaries of individual scientific disciplines.[40] Unless this can be effectively applied, disciplinary boundaries become impermeable to intellectual synergy and hinder the enlightenment we find only at the interfaces between disciplines.

Improving our understanding of the ecological integrity – environmental goods – human health linkages will aid us as we seek to develop an acceptable standard of living for many more people, while we ensure that the ecological pillars, which support our society and industries, are protected and remain sustainable.[2–4] We must successfully integrate social and natural systems on a local scale, while understanding the larger scale ramifications and consequences of decisions on a local, national and trans-national scale. Our knowledge of the biosphere and our connections with it must be adequate to inform policy and decision-making.

Harmful impacts of pollutants on ecological integrity are complex, often subtle, and may go undetected until the 'fire alarm' rings. However, effective risk evaluation and forecasting of impact requires a capability for interpreting and integrating complex environmental information that is currently lacking (International Risk Governance Council, 2010, http://irgc.org/IMG/pdf/irge_ER_final_07jan_web.pdf). A practical solution to this problem lies in a 'whole systems' approach to the development of early-warning strategies for assessing risk to aquatic ecosystem function and health, which effectively integrate physical, chemical and biological processes. To achieve this, we need a novel approach and can use existing environmental data and diagnostic tools for the detection of 'distress signals' and evaluation of 'ecosystem health status' in novel ways.[4,5] Integration of this complex information can be achieved using whole system-based simulation modelling; this will also facilitate the targeting of current 'knowledge gaps' and how to fill them.

Consequently, a hierarchy of overlapping computational models will be needed, each capable of interacting with the others.[7,8] Development of these models will have to be hypothesis-driven and will, of necessity, integrate physical and chemical processes with ecosystem function using a systems-biology approach. The aim here will be the development of an affordable pre-operational predictive toolbox to aid decision-making in a systems-based holistic way, for advising science-based environmental policy formulation. Achievement of this aim will require a critical interdisciplinary evaluation of the existing physical and chemical parameters, diagnostic and prognostic biomarkers of health status, and indices of biodiversity/ecosystem integrity currently used as indicators of biological and ecological damage.[5,11,26,76,81,98,125] This process will also include an assessment of the ecological resilience of critical habitats, like river basins and coastal zones.[10,13,16]

6 Conclusions and Recommendations

We need an integrated explanatory framework for adverse changes in whole systems, from the physical environment to cells, to animals/plants, to human health and ecosystems. Frameworks for the evaluation of 'health of the environment' and prediction of ecosystem consequences resulting from future environmental events should encompass integrating physical transport and bioavailability of contaminants and pathogens with the reactions of biomarkers of 'harmful effect' for cellular and physiological processes with whole-system function, through conceptual and numerical modelling.[76,78,81,125,126] These frameworks are urgently needed to synthesise complex information on environmental chemistry and injurious effects of pollutants into predicted harmful impact on the health of sentinel animals and ecosystems. Predictive validity is an important component of assessing ecological, human and socio-economic risk.

To achieve the above goal, we must fill the following research needs or gaps if we are to develop a holistic understanding of the interconnections between functionality in individual organisms and ecological assemblages, as well as possible human and societal consequences:[4,77]

1. Characterise the key physical and chemical factors involved in determining contaminant binding to natural particulates, transport and biological availability;
2. Link environmental stress effects at the molecular and cellular levels with higher-level ecological and human health consequences, in order to provide a biological basis for understanding similarities and differences of vulnerable subsets of sentinel organisms with ecosystems;
3. Identify effective research interfaces that enable discovery of the consequences of environmental stressors throughout the hierarchical levels of biological organisation. For example, how can environmental mathematicians and computer modellers provide insight into the links between molecular- or cellular-level responses and higher-level individual animal health, ecological consequences and impacts on human health?
4. Develop a mechanistic understanding of intracellular transport, chemistry and biotransformation of pollutants within food webs. How can such understanding help us identify the human health risks?
5. How do interconnections between individual organism health and ecological integrity change as a function of the rate of ecosystem degradation or loss of assimilative capacity?
6. Understand the assimilative capacity of vulnerable ecosystems. To what extent must ecosystem assimilative capacity degrade before environmental goods (*e.g.* water quality and fisheries) are affected detrimentally?
7. Understand the extent to which boundaries of specific ecosystems overlap and thus interconnect with human populations. Do such ecosystem boundaries impact different human populations in different ways; and can we use indices of adverse impact on ecosystem function as potential early indicators of harmful interactions with human health?
8. Develop mathematical and simulation models at various levels of biological complexity to address the complexity of ecological integrity interconnections;
9. Identify the role of science in decision-making, and enhance the integration of science into the process while recognising its inherent limitations;

10. Develop strategies for improving the infrastructure for interdisciplinary research in order to promote collaboration between physical scientists, biologists, ecologists and policy makers;
11. Develop strategies for improving communication among scientists, industrialists, policy-makers and the public to reduce mistrust and misunderstanding.

The interrelated character of the above issues and questions requires a holistic approach if they are to be effectively addressed.[3,14] Hence, any proposed integrated scientific and environmental health management strategy must be truly interdisciplinary if an effective capability for risk assessment and prediction is to be developed in relation to resource sustainability. The process of estimating risk associated with the possibility of future events in the coastal zone has two principal elements:

1. A capacity to predict environmental impacts which, when related to dynamic processes, typically involves the use of simulation models in environmental forecasting of natural and man-induced events that have public health consequences;
2. The calculation of risk itself, which requires a probabilistic analysis of predicted events using standard procedures, such as Monte Carlo simulations, Markov Monte Carlo methods, dependency bounds analysis and 'fuzzy logic' theory.[13,16]

Judgement of the acceptability of risk is not an issue for scientific research alone, as it involves economic, social and other issues. These elements all need to be integrated in order to develop prognostic risk models, which are suitable for operational application. Precaution, or the reduction of risk, has important economic components related to the value of the resources to be protected and the costs of increasing safety margins. Thus, risk assessment is intrinsically a cross-disciplinary issue.

Programmes for environmentally sustainable management of coastal areas must engender a move away from the traditional sector-by-sector management practice to a cross-sectoral approach. This practical and philosophical transformation is essential, in order to deliver the capacity to anticipate and minimise risks by improving the definition in space and time of master variables that determine change in environmental quality. Risk is sometimes greater than has been estimated (*i.e.* hyperconservative) because of the incompleteness of data (*e.g.* point samples compared with airborne or satellite remote-sensed data), or marked heterogeneities in key variables (*e.g.* accumulation of toxic contaminants at an interface), or non-linear processes (*e.g.* toxicity thresholds), where small changes in concentrations may have marked harmful effects. What is needed is an effective conservative methodology, such as 'dependency bounds analysis', that makes no assumptions unwarranted by empirical evidence.[26,111]

Effective coastal and estuarine management has the prerequisite of a sound foundation of scientific research and understanding of environmental processes. This in turn requires collaboration between environmental managers (*i.e.* in national and international environmental protection agencies) and research centres with the strategic capability to understand processes of change and the impact of human activities upon them. In addition, it is important that the various governments, regional organisations (*e.g.* EU) and global UN organisations (*e.g.* GEF – Global Environment Facility; UNDP – United Nations Development Programme; UNEP – United Nations Environment Programme; IOC-UNESCO – Inter-Governmental Oceanographic Commission, United Nations Educational, Scientific and Cultural Organisation; IMO – International Maritime Organisation; WMO – World Meteorological Organisation; WHO – World Health Organisation; and UNIDO – United Nations Industrial Development Organisation), which have interests in ICZM, harmonise their activities in order to effect synergies in this global activity.

It is also considered crucial that research programmes address broader socio-economic issues, involving people-orientated environmental health-related problems, such as human population pressure. Unfortunately, there is still a relative dearth of substantial epidemiological data that would permit a comprehensive understanding of possible causal links between human and ecosystem health (see Figure 2; *Millennium Ecosystem Assessment*, 2005, http://www.millenniumassessment.org/en/index.aspx; and World Health Organisation, http://www.who.int/topics/environmental_health/en/). We must aspire to gain a better understanding of how we interact with our environment and what the consequences will be for sustainability of environmental goods and human health. This goal requires that we have the necessary multidisciplinary capacity capable of responding in an interdisciplinary mode to resolve problems that are intrinsically interfacial in character. By effectively identifying and interconnecting the interdisciplinary elements, we will see the emergence of new ways of solving problems in what at present are seemingly unrelated areas of environment and human health.

A major reason for the development of complexity science, and its use in ecotoxicology and environmental toxicology, is to gain a realistic insight into the limits of reductionism as a very successful universal problem-solving approach (see Figure 3).[117,127] Complex biological and ecological processes generate counter-intuitive, seemingly acausal behaviour that is full of novelty.[85,117,128] Trying to understand the behaviour of a complex adaptive (or dynamic) system, such as an organism, population, ecosystem or economy, by a reductionist approach often irretrievably destroys the inherent nature of the problem.[117] Recognition that a system is complex is specifically subjective, not an objective property of an isolated system. However, it can become objective, once the investigative formalism takes into account the larger system with which the target system itself interacts.[117,127]

There must also be a wider recognition that ICZM and ecotoxicology/environmental toxicology are dealing with complex adaptive systems.[88,117]

Environment & Health – Integration & Forecasting

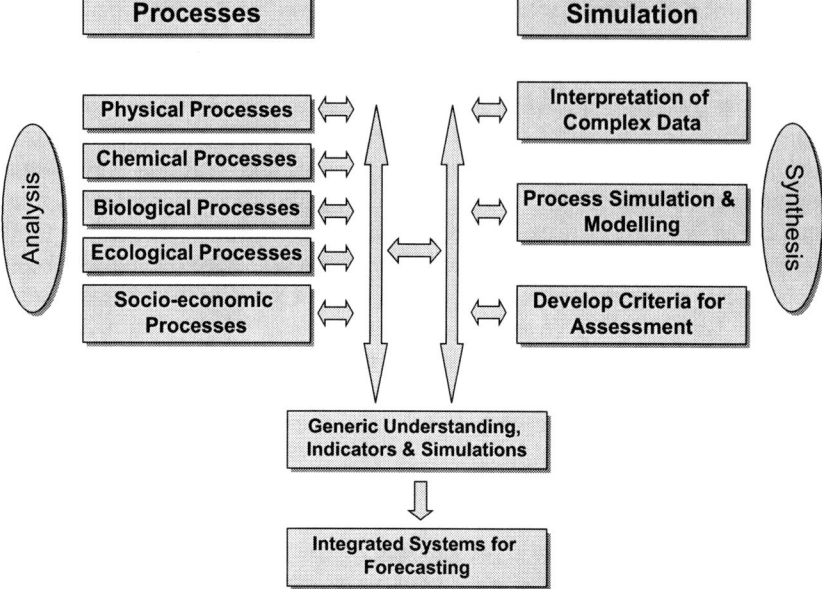

Figure 3 Proposed co-evolutionary analytical and systems approach for integration of environmental, health and socio-economic data. (Adapted from Moore et al.)[26]

Consequently, there needs to be a rapid acquisition of the new methods of 'complexity science' and incorporation of these into the scientific components of integrated environmental management programmes (see Figure 3). Therefore, a rational way forward can be conceived if an integrated multidisciplinary approach to the environmental impact of various natural and anthropogenic stressors is adopted, as follows:

- Development of conceptual frameworks and process-simulation models based on an improved mechanistic understanding of contaminant uptake, biotransformation, toxicity and impact within the biological organisational hierarchy (see Figure 2);[78,88,91,118,128–131]
- Application to the interdisciplinary challenge of modelling processes in environmental pollution and impact as a complex adaptive system;[117]
- Adoption of a broad view of the current and predicted future problems in environmental management, that incorporates both moderate reductionist and synthetic approaches;
- Collaboration on areas to include, among others, remote/satellite surveillance, risk assessment, interpretation of complex information and predictive modelling;

- Precautionary anticipation of novel environmental hazards (*e.g.* from biotechnology and molecular nanotechnology);
- Prevention of haphazard coastal domestic and industrial development; stop illegal disposal of waste and introduce effective waste management and cleaner industrial production;
- Improved communication with politicians, industrialists and environmental managers concerning the long-term consequences of pollution and environmental degradation;
- Improved public understanding of science and risk;
- Mitigation of the impacts of human population expansion in vulnerable coastal areas through better education, effective advice about family planning and its free availability, coupled with good governance (*e.g.* sustainable agriculture, industrial and urban development and effective waste management).

Finally, it is worth highlighting two generic issues which impinge on the provision of scientific evidence to ensure that effective policies are developed to address the challenges we face. These are our failure to reach out from disciplinary silos to work closely with other disciplines, and the weakness of communication between the scientific community and the policy community, and *vice versa*.[124] For example, marine environmental scientists and policy-makers are usually unaware of the intimate links between changes in climate and marine environmental quality, human health and wellbeing. This is because they seldom meet in common *fora*. Added to this is the need for experts in other fields to work with environmental and medical scientists and policy makers to consider social, economic and legal aspects that influence the interconnections between the marine environment and human health. Most governments also struggle to ensure that their various departments work together. This is easily illustrated by questioning how often officials from departments of health meet with departments of environment officials, and how they interact with departments responsible for climate change, food security, industrial development, transport and the like. The answer is clearly 'not very often, if at all'. Secondly, concerning the underpinning of policies with scientific evidence, a process has been identified by which this is supposed to occur – the so-called 'science–policy cycle'. However, in practice this often leads to unsatisfactory policies that result in unwanted, unexpected consequences. Scientists are often accused of being poor at explaining their findings. However, the science-policy interface is bi-directional, and policy makers are often ill-equipped to understand basic scientific messages in an appropriate context. Worse still, as was alluded to earlier, numerous other important actors such as social scientists, economists, political scientists and lawyers are often excluded from the process. It is also clear that many policy developments occur serendipitously as a result of a particularly strong advocate engaging with a proactive politician or policymaker. There may be little that can or should be done about this, but we should at least be aware of the phenomenon.

Establishing iterative dialogues, making use of the expertise of all of the forementioned groups would be a valuable step forward in improving the use of scientific evidence in policymaking, nowhere more so than in relation to the issue of the marine environment and human health.

References

1. D. M. Roodman, *The Natural Wealth of Nations: Harnessing the Market for the Environment*, The Worldwatch Environmental Alert Series, W. W. Norton and Co., New York, 1998.
2. A. M. Torres and C. A. Monteiro, *J. Epidemiol. Community Health*, 2002, **56**, 82.
3. R. T. Di Giulio and W. B. Benson, *Interconnections between Human Health and Ecological Integrity*, Society of Environmental Toxicology and Chemistry (SETAC), Pensacola, FL, 2002.
4. L. Westra, K. Bosselmann and R. Westra, *Reconciling Human Existence with Ecological Integrity - Science, Ethics, Economics and Law*, Earthscan, London, 2008.
5. R. E. Bowen and M. H. Depledge, *Mar. Pollut. Bull.*, 2006, **53**, 541.
6. L. E. Fleming, K. Broad, A. Clement, E. Dewailly, S. Elmir, A. Knap, S. A. Pomponi, S. Smith, H. Solo-Gabriele and P. Walsh, *Mar. Pollut. Bull.*, 2006, **53**, 545.
7. J. I. Allen, ch.1 of this volume, 2011.
8. L. Burke, Y. Kura, K. Kassem, M. Spalding and C. Revenga, *Pilot Analysis of Global Ecosystems: Coastal Ecosystems Technical Report*, World Resources Institute, Washington, DC, 2000.
9. D. Bryant, E. Rodenburg, T. Cox and D. Nielsen, *Coastlines at Risk: An Index of Potential Development-Related Threats to Coastal Ecosystems*, WRI Indicator Brief, World Resources Institute, Washington, DC, 1995.
10. L. F. Cassar, *Integrated Coastal Area Management*, United Nations Industrial Development Organisation, Vienna, 2001.
11. T. S. Galloway, *Mar. Pollut. Bull.*, 2006, **53**, 606.
12. N. Hardman-Mountford and J. M. McGlade, in *The Gulf of Guinea Ecosystem*, ed. J. M. McGlade, K. Korangteng, P. Cury and N. Hardman-Mountford, Elsevier, Amsterdam, 2002, 49.
13. M. N. Moore and Z. Csizer, in *Integrated Coastal Area Management*, ed. L. F. Cassar, UNIDO, Vienna, 2001, 161.
14. E. C. D. Todd, *Mar. Pollut. Bull.*, 2006, **53**, 569.
15. P. J. Walsh, S. L. Smith, L. E. Fleming, H. M. Solo-Gabrielle and W. H. Gerwick, *Oceans and Human Health. Risks and Remedies from the Seas*, Elsevier, 2008.
16. J. M. McGlade, in *Science and Integrated Coastal Zone Management*, Dahlem Conference 86, ed. B. von Bodungen and R. K.Turner, Dahlem University Press, 2001, 349.
17. J. Quarrie, *Earth Summit '92*, Regency Press, London, 1992.
18. L. Mee, *AMBIO*, 1992, **21**, 278.
19. Population Reports, *Population and the Environment: the Global Challenge, Special Topics*, Johns Hopkins School of Public Health, Baltimore,

MD, 2000, vol. 28, No. 3, series M, No. 15, http://info.k4health.org/pr/m15/m15print.shtml.
20. H. L. Windom, *Mar. Pollut. Bull.*, 1992, **25**, 32.
21. R. Costanza, R. d'Arge, R. de Groot, S. Farber, M. Grasso, B. Hannon, K. Limburg, S. Naeem, R. V. O'Neill, J. Paruelo, R. G. Raskin, P. Sutton and M. van den Belt, *Nature*, 1997, **387**, 253.
22. United Nations Environment Programme (UNEP), *Change and Challenge: An Introduction to the West and Central African Action Plan*, UNEP, Nairobi, 1991.
23. United Nations Population Division, *World Urbanization Prospects: The 1999 Revision*, UN, New York, 1999.
24. World Bank, *Africa: A Framework for Integrated Coastal Zone Management*, World Bank, Washington, DC, 1995.
25. M. Maso and E. Garces, *Mar. Pollut. Bull.*, 2006, **53**, 620.
26. M. N. Moore, M. H. Depledge, J. W. Readman and P. Leonard, *Mutat. Res.*, 2004, **552**, 247.
27. J. C. Orr, V. J. Fabry, O. Aumont, L. Bopp, S. C. Doney, R. A. Feely, A. Gnanadesikan, N. Gruber, A. Ishida, F. Joos, R. M. Key, K. Lindsay, E. Maier-Reimer, R. Matear, P. Monfray, A. Mouchet, R. G. Najjar, G.-K. Plattner, K. B. Rodgers, C. L. Sabine, J. L. Sarmiento, R. Schlitzer, R. D. Slater, I. J. Totterdell, M.-F. Weirig, Y. Yamanaka and A. Yool, *Nature*, 2005, **437**, 681.
28. M. H. Depledge and W. J. Bird, *Mar. Pollut. Bull.*, 2009, **58**, 947.
29. M. White, A. Smith, K. Humphryes, D. Snelling and M. H. Depledge, *J. Environ. Psychol.*, 2010, **30**, 482.
30. A. De Moor and P. Calamai, *Subsidizing Unsustainable Development: Undermining the Earth with Public Funds*, Earth Council, Toronto, 1997, No. 6.
31. Food and Agriculture Organisation of the United Nations (FAO), *The State of World Fisheries and Aquaculture*, FAO, 1998. Rome, 1999.
32. A. P. McGinn, *Safeguarding the Health of the Oceans*, Worldwatch Paper, 1999, No 145.
33. P. M. Kris-Etherton, W. S. Harris and L. J. Appel, *Circulation*, 2002, **106**, 2747.
34. C. L. Loprinzi, R. Levitt, D. L. Barton, J. A. Sloan, P. J. Atherton, D. J. Smith, S. R. Dakhil, D. F. Moore, J. E. Krook, K. M. Rowland, M. A. Mazurczak, A. R. Berg and G. P. Kim, *Cancer*, 2005, **104**, 176.
35. D. Hinrichsen, *Coastal Waters of the World: Trends, Threats and Strategies*, Island Press, Washington, DC, 1998.
36. United Nations Environment Programme (UNEP), *State of the Marine Environment in the South-East Pacific Region*, UNEP, Bogota, 1988.
37. World Bank, *Africa: A Framework for Integrated Coastal Zone Management*, World Bank, Washington, DC, 1995.
38. M. H. Depledge, Rapid assessment of marine pollution (RAMP), in *Proceedings of the Vietnam-UK Joint Workshop on Marine Pollution Assessment, Hanoi, June 8, 2000*, National Centre for Natural Science and Technology of Vietnam, Hanoi, Vietnam, 2000, 5.

39. E. Roth and H. Rosenthal, *Mar. Pollut. Bull.*, 2006, **53**, 599.
40. M. N. Moore and P. D. Kempton, *Environ. Health*, 2009, **8**(Suppl 1), S1, doi: 10.1186/1476-069X-8-S1-S1.
41. Population Reference Bureau (PRB), *World Population Data Sheet*, Population Reference Bureau, Washington, DC, 2000.
42. The Royal Commission on Environmental Pollution, *Demographic Change and the Environment*, Royal Commission on Environmental Pollution, UK Crown Copyright, London, 2011.
43. A. B. A. Boxall, A. Hardy, S. Beulke, T. Boucard, L. Burgin, P. D. Falloon, P. M. Haygarth, T. Hutchinson, R. S. Kovats, G. Leonardi, L. S. Levy, G. Nichols, S. A. Parsons, L. Potts, D. Stone, E. Topp, D. B. Turley, K. Walsh, E. M. H. Wellington and R. J. Williams, *Environ. Health Perspect.*, 2009, **117**, 508.
44. J. W. Readman, R. F. C. Mantoura and M. M. Rhead, *Fresenius Z. Anal. Chim.*, 1984, **319**, 126.
45. R. C. Thompson, C. J. Moore, F. S. vom Saal and S. H. Swan, *Philos. Trans R. Soc. London, Ser. B*, 2009, **364**, 2153.
46. D. Thomas, B. Gustafsson and Ö. Gustafsson, *Environ. Sci. Technol.*, 2000, **34**, 5144.
47. M. St. J. Warne and D. W. Hawker, *Ecotox. Environ. Safety*, 1995, **31**, 23.
48. J. L. Zhou, T. W. Fileman, S. Evans, P. Donkin, J. W. Readman, R. F. C. Mantoura and S. Rowland, *Sci. Total Environ.*, 1999, **244**, 305.
49. H. I. Berry, K. Bowen and T. Kjellstrom, *Int. J. Public, Health*, 2010, **55**, 123.
50. J. L. Zhou, T. W. Fileman, S. V. Evans, P. Donkin, C. A. Llewellyn, J. W. Readman, R. F. C. Mantoura and S. J. Rowland, *Mar. Pollut. Bull.*, 1998, **36**, 597.
51. L. Ababouch, *Mar. Pollut. Bull.*, 2006, **53**, 561.
52. K. Davidson and E. Bresnan, *Environ. Health*, 2009, **8**(Suppl 1), S12, doi: 10.1186/1476-069X-8-S1-S12.
53. H. Toyofuku, *Mar. Pollut. Bull.*, 2006, **53**, 579.
54. T. Yasuda and R. E. Bowen, *Mar. Pollut. Bull.*, 2006, **53**, 640.
55. F. Feldhusen, *Microbes Infect.*, 2000, **2**, 1651.
56. Defra, *R&D Newsletter Agriculture and the Environment*, 2004, **12**, 8.
57. A. Rzezutka and N. Cook, *FEMS Microbiol. Rev.*, 2004, **28**, 441.
58. M. N. Moore, *Environ. Int.*, 2006, **32**, 967.
59. R. Owen and M. H. Depledge, *Mar. Pollut. Bull.*, 2005, **50**, 609.
60. R. D. Handy, T. B. Henry, T. M. Scown, B. D. Johnston and C. R. Tyler, *Ecotoxicology*, 2008, **17**, 396.
61. R. D. Handy, F. von der Kammer, J. R. Lead, M. Hassellov, R. Owen and M. Crane, *Ecotoxicology*, 2008, **17**, 287.
62. M. N. Moore, J. A. Readman, J. W. Readman, D. M. Lowe, P. E. Frickers and A. Beesley, *Nanotoxicology*, 2009, **3**, 40.
63. J. Panyam and V. Labhasetwar, *Adv. Drug. Delivery Rev.*, 2003, **55**, 329.
64. J. Panyam, S. K. Sahoo, S. Prabha, T. Bargar and V. Labhasetwar, *Int. J. Pharm.*, 2003, **262**, 1.
65. L. Pelkmans and A. Helenius, *Traffic*, 2002, **3**, 311.

66. V. Stone and K. Donaldson, *Nature Nanotechnol.*, 2006, **1**, 23.
67. V. Stone, B. Nowack, A. Baun, N. van den Brink, F. von der Kammer, M. Dusinska, R. Handy, S. Hankin, M. Hassellöv, E. Joner and T. F. Fernandes, *Sci Total Environ.*, 2010, **408**, 1745.
68. D. M. Brown, M. R. Wilson, W. MacNee, V. Stone and K. Donaldson, *Toxicol. Appl. Pharmacol.*, 2001, **175**, 191.
69. Royal Society and Royal Academy of Engineering, *Nanoscience and Nanotechnologies: Opportunities and Uncertainties*, RS Policy Document 19/04, The Royal Society, London, 2004.
70. P. H. M. Hoet, I. Brüske-Hohlfeld and O. V. Salata, *J. Nanobiotechnol.*, 2004, **2**, 12, doi: 10.1186/1477-3155-2-12.
71. M. Hassellöv, J. W. Readman, J. Ranville and K. Tiede, *Ecotoxicol.*, 2008, **17**, 344.
72. A. T. C. Bosveld, P. A. F. de Bie, N. W. van den Brink, H. Jongepier and A. V. Klomp, *Chemosphere*, 2002, **49**, 75.
73. A. Baun, N. B. Hartmann, K. D. Grieger and S. F Hansen, *J. Environ. Monit.*, 2009, **11**, 1774.
74. M. H. Depledge, J. J. Amaral-Mendes, B. Daniel, R. S. Halbrook, P. Kloepper-Sams, M. N. Moore and D. P. Peakall, in *Biomarkers – Research and Application in the Assessment of Environmental Health*, ed. D. G. Peakall and L. R. Shugart, Springer, Berlin, Heidelberg, 1993, 15.
75. T. S. Galloway, R. C. Sanger, K. L. Smith, G. Fillmann, J. W. Readman, T. E. Ford and M. H. Depledge, *Environ. Sci. Technol.*, 2002, **36**, 2219.
76. T. S. Galloway, J. A., Hagger, D. Lowe, D. R. P Leonard, R. Owen and M. B. Jones, *Can Biomarkers Measure the Environmental Health of Estuaries?* ICES CM, 2007, 1:01.
77. G. W. Winston, S. M. Adams, W. H. Benson, L. E. Gray, H. S. Matthews, M. N. Moore and S. Safe, in *Interconnections between Human Health and Ecological Integrity*, ed. R. T. Di Giulio and W. B. Benson, Society of Environmental Toxicology and Chemistry (SETAC), Pensacola, FL, 2002, 43.
78. J. I. Allen and M. N. Moore, *Mar. Environ. Res.*, 2004, **58**, 227.
79. J. W. Readman, E. Kadar, J. A. J. Readman and C. Guitart, ch. 3 in this volume, 2011.
80. R. Owen, M. H. Depledge, J. A. Hagger, M. B. Jones and T. S. Galloway, *Mar. Pollut. Bull.*, 2008, **56**, 613.
81. J. A. Hagger, M. B. Jones, D. Lowe, D. R. P. Leonard, R. Owen and T. S. Galloway, *Mar. Pollut. Bull.*, 2008, **56**, 1111.
82. K. Hylland, *Mar. Pollut. Bull.*, 2006, **53**, 614.
83. A. J. Lawrence, A. Arukwe, M. N. Moore, M. Sayer and J. Thain, in *Effects of Pollution on Fish – Molecular Effects and Population Responses*, ed. A. Lawrence and K. Hemingway, Blackwell Science Ltd., Oxford, 2003, 83.
84. M. N. Moore, J. I. Allen and P. J. Somerfield, *Mar. Environ. Res.*, 2006, **62**(Suppl. 1), S420.
85. C. V. Howard, *The Ecologist*, 1997, **27**, 192.
86. A. Rtenkamp and R. Altenburger, *Sci. Total Environ.*, 1998, **221**, 59.

87. B. Kurelec, The genotoxic disease syndrome, *Mar. Environ. Res.*, 1993, **35**, 341.
88. M. N. Moore, *Aquatic Toxicol.*, 2002, **59**, 1.
89. A. Murdoch, K. L. E. Kaiser, M. E. Comba and M. Neilson, *Sci. Total. Environ.*, 1994, **158**, 113.
90. F. Smedes, *Int. J. Environ. Anal. Chem.*, 1994, **57**, 215.
91. J. I. Allen and A. McVeigh, *J. Mol. Histol.*, 2004, **35**, 697.
92. M. N. Moore and D. Noble, *J. Mol. Histol.*, 2004, **35**, 655.
93. D. R. P. Leonard and G. J. Hunt, *J. Soc. Radiol. Protect.*, 1995, **5**, 129.
94. D. R. P. Leonard and R. J. Pentreath, *Mar. Biol.*, 1981, **63**, 67.
95. D. R. P. Leonard, K. J. Mondon and M. G. Segal, *J. Soc. Radiol. Protect.*, 1993, **13**, 43.
96. P. G. Wells, M. H. Depledge, J. N. Butler, J. J. Manock and A. H. Knap, *Mar. Pollut. Bull.*, 2001, **42**, 799.
97. D. L. Archer, *Int. J. Food Microbiol.*, 2004, **90**, 127.
98. R. E. Bowen and M. H. Depledge, *Mar. Pollut. Bull.*, 2006, **53**, 631.
99. H. M. Crews and A. B. Hanley, *Biomarkers in Food Chemical Risk Assessment*, The Royal Society of Chemistry, London, 1995.
100. M. Hahn, *Sci. Total Environ.*, 2002, **289**, 49.
101. J. Zhou, X.-S. Zhu and Z.-H. Cai, *J. Mar. Animals Ecol.*, 2009, **2**, 7.
102. J. Rice, *Ocean Coastal Manag.*, 2003, **46**, 235.
103. EU Water Framework Directive, *Directive 2000/60.EC of the European Parliament and of the Council for the Community Action in the Field of Water Policy*, 2000, http://ec.europa.eu/environment/water/water-framework.
104. D. d'A. Laffoley, J. Burt, P. Gilland, J. Baxter, D. W. Connor, M. Hill, J. Breen, M. Vincent and E. Maltby, *Adopting an Ecosystem Approach for the Improved Stewardship of the Maritime Environment*, English Nature Research Reports, 2003, No 538.
105. D. R. P. Leonard, *Mar. Environ. Res.*, 2002, **54**, 209.
106. M. H. Depledge and Z. Billinghurst, *Mar. Pollut. Bull.*, 1999, **39**, 32.
107. S. Jobling and R. Owen, in *Late Lessons from Early Warnings*, ed. D. Gee, European Environment Agency, 2011.
108. A. M. Mayer, K. B. Glaser, C. Cuevas, R. S. Jacobs, W. Kem, R. D. Little, J. M. McIntosh, D. J. Newman, B. C. and D. E. Shuster, *Trends Pharmacol. Sci.*, 2010, **31**, 255.
109. J. Thompson Coon, K. Boddy, K. Stein, R. Whear, J. Barton and M. H. Depledge, *Environ. Sci. Technol.*, 2011, in press.
110. K. M. Brander, *Proc. Natl. Acad. Sci. U. S. A.*, 2007, **104**, 19709.
111. S. Ferson and T. F. Long, in *Environmental Toxicology and Risk Assessment,* ed. J. S. Hughes, ASTM, Philadelphia, PA, 1995, 97.
112. A. N. Jha, V. V. Cheung, M. E. Foulkes, H. J. Hill and M. H. Depledge, *Mutat. Res.*, 2000, **464**, 213.
113. J. A. Caswell, *Mar. Pollut. Bull.*, 2006, **53**, 650.
114. J. de Goede, J. M. Geleijnse, J. M. A. Boer, D. Kromhout and W. M. M. Verschuren, *J. Nutr.*, 2010, **140**, 1023.

115. H. H. Jensen, *Mar. Pollut. Bull.*, 2006, **53**, 591.
116. S. Brenner, in *The Limits of Reductionism in Biology*, ed. G. R. Bock and J. A. Goode, Novartis Foundation Symposium, John Wiley, London, 1998, **213**, 106.
117. S. A. Kauffman, *The Origins of Order*, Oxford University Press, New York, 1993.
118. D. Noble, J. Levin and W. Scott, *Drug Discovery Today*, 1999, **4**, 10.
119. A. McVeigh, J. I. Allen, M. N. Moore, P. Dyke and D. Noble, *Mar. Environ. Res.*, 2004, **58**, 821.
120. D. Noble, *Nature Rev. Mol. Cell Biol.*, 2002, **3**, 460.
121. J. B. Wiener and M. D. Rogers, *J. Risk Res.*, 2002, **5**, 317.
122. C. Vincent, H. Heinrich, A. Edwards, K. Nygaard and K. Haythornthwaite, *Guidance on Typology, Reference Conditions and Classification Systems for Transitional and Coastal Waters*, CIS Working Group 2.4 (COAST), Common Implementation Strategy of the Water Framework Directive, European Commission, 2002.
123. Environment Agency, *The Water Framework Directive: Guiding Principles on the Technical Requirements*, Environment Agency, UK, 2002.
124. M. H. Depledge, *Ocean Coastal Manag.*, 2009, **52**, 336.
125. D. R. P. Leonard, K. R. Clark, P. J. Somerfield and R. Warwick, *J. Environ. Manag.*, 2006, **78**, 52.
126. J. A. Haggar, M. B. Jones, D. R. P. Leonard, R. Owen and T. S. Galloway, *Integr. Environ. Assess. Manag.*, 2006, **2**, 312.
127. J. L. Casti, *Complexification: Explaining a Paradoxical World through the Science of Surprise*, Abacus, London, 1994.
128. F. Kanzawa, K. Nishio, K. Fukuoka, M. Fukuda, T. Kunimoto and N. Saijo, *Int. J. Cancer*, 1997, **71**, 311.
129. J. I. Allen, J. Blackford, J. Holt, R. Proctor, M. Ashworth and J. Siddorn, *Sarsia*, 2001, **86**, 423.
130. M. N. Moore, *Mar. Environ. Res.*, 2010, **69**, S37.
131. M. N. Moore and J. I. Allen, *Mar. Environ. Res.*, 2002, **54**, 579.
132. Millennium Ecosystem Assessment, 2005, http://www.millenniumassessment.org/en/index.aspx.
133. World Health Organisation, 2005, http://www.who.int/topics/environmental_health/en/.

Subject Index

A
Acidification of the oceans, 8
Action Plan on Environment and Health, 149
Additives, 76
Agricultural runoff, 10
Algal toxins, 116, 138
Amnesic shellfish poisoning (ASP), 13, 101–102, 110
Amoebic dystentry, 30
Anaerobic bacteria, 39
Antibiotics, 12, 76
Antioxidant substances, 77
Aquaculture, 2, 7–9, 13, 138
Azaspiracid poisoning (AZP), 111

B
Bacteria, 28–29, 31
Bacterial pathogens, 30
Ballast water, 119
Bathing Water
  Directive, 147
  quality, 9
BEACH Act, 33
Beach
  closures, 37
  regulation, 56–57
  sand, 37
  water profiles, 58
Beaches, 34, 37, 56
Bioaccumulative effects, 11
Bioavailability, 86
Biodiversity, 2–3, 8
Bioinformatics, 19
Biological systems modelling, 146
Biomarkers, 19

Biotoxin
  monitoring, 113
  -producing phytoplankton, 109, 114
Biotoxins, 99
Blue
  green algae, 98
  gym, 14, 144
  effect, 133, 144

C
Carcinogens, 75, 85
Carcinogenic, mutagenic and reprotoxic chemicals, 150
Carcinogenicity, 117
Cargo vessel ballast tanks, 119
Chemical warfare agents, 78
Cholera, 9, 27, 31, 54–55
Ciguatera fish poisoning (CFP), 105, 116
Ciguatoxin, 106
Clean Water Act, 33
Climate
  change, 2–4, 86–88, 120, 137, 139
  impacts, 88
  impacts, 16
Coastal
  biodiversity, 134
  populations, 7
Cocktails of priority pollutants, 85
Coliforms, 32–33, 39
Coliphages, 39
Collapse of fisheries, 129
Complex adaptive systems, 155
Complexity science, 155–156
Computational models, 142, 151

Congress on Evolutionary
  Computation, 149
Contaminated seafood, 26
Critical coastal ecosystems, 145
Culture-based methods, 38
Cyanobacteria, 98, 106, 116
  toxins, 138

D
Deepwater Horizon oil spill, 70
Desalination plants, 79
Detergent residues, 78
Diarrhetic shellfish
  poisoning (DSP), 13, 102, 104, 111
    outbreaks, 103
    toxin production, 103
Diatoms, 108, 113
Dinoflagellates, 107, 112
Dioxins, 76
Discolouration of the sea, 98
Disinfection byproducts, 79
Driver-Pressure-State-Impact-
  Response-Model (DPSIR), 5–6

E
Early warning methodologies, 117
Earth Summit Conference in Rio de
  Janeiro, 130
Ecological indicators, 143
Ecosystem
  approach to management, 15
  integrity, 135
  services, 134–135
Ecotoxicology, 134, 136
El Niño Southern Oscillation
  (ENSO), 4, 120
Emerging contaminants, 74, 78
Endocrine
  disrupters, 86
  disruption, 11, 143
Engineered
  nanoparticle (ENP)
    production, 80
  nanoparticles, 83
  plastics, 75

Enteric
  pathogens, 35, 43
  viruses, 11, 29, 40, 56
Enterococci, 33–35
Environmental
  goods and services, 129
  impact, 144
  model, 20
  policy framework, 136
  prognostics, 18
  Protection Agency, 34
  quality, 134
Epidemiological
  data, 151
  studies, 33, 36, 44–45, 52–53
Escherichia coli, 32
EU Environment and Health
  ERANET, 149
Eutrophication, 8–9, 13, 97–98

F
Farmed
  fish, 107
  salmon, 113
Fecal
  contamination, 42
  indicator, 29, 32–33, 42, 56
    bacteria (FIB), 28–29, 32–33, 56
  indicators, 40
  pollution, 39, 58
Fish
  stocks, 99
  vectored illness, 105
Foaming of the sea, 107
Food
  hygiene regulations, 114
  Standards Agency for
    Scotland, 114
Forecasting, 21
Fungal pathogens, 30

G
Gastroenteritis, 11
Gastrointestinal (GI) disease, 26
Genes, 31
Genetic fingerprinting, 41

Globalisation, 139
Goods and services, 1

H
Habitat degradation, 129
Harm to human health, 99
Harmful
   algae, 98
   algal bloom (HAB), 9–10, 12, 95, 98, 108
      operational forecast (HAB-OF), 119
      programme, 100
      in UK coastal waters, 108
   phytoplankton, 98, 106, 111
Hazard Action Critical Control Points (HACCP), 116
Health
   challenges, 150
   policy priorities, 149
Healthy Beaches programs, 34
Human
   fecal pollution, 36–38
   health and wellbeing, 10

I
Ideal indicator, 38
Indicator, 28, 38, 58
   microbes, 58
Indicators of
   ecological integrity, 151
   human fecal pollution, 40
Indices of environmental impact, 144
Indigenous pathogens, 30
Industrial emerging contaminants, 75
Integrated
   coastal zone management (ICZM), 148–149
   environmental management, 156
Interconnections in marine environment and health, 131
International
   Centre for Science and High Technology, 136
   Risk Governance Council, 150
Invasive species, 139
Irrigation of crops, 139

K
Karenia mikimotoi, 111
Key science challenges, 135

L
Life Cycle Assessment (LCA), 83
Lipophilic shellfish toxin (LST), 103

M
Marine
   ecosystem, 3, 8
   Strategy Framework Directive, 5, 15, 73
MARPOL convention, 85
Mathematical models, 54–55
   for cholera, 54
Mercury, 3, 12, 68–69
Methicillin-resistant *S. aureus* (MRSA), 30
Methylmercury, 11, 69
Microbial
   infection, 69
   measures of water quality, 45
   monitoring, 57
   pathogens, 10–11, 17, 25
   quality, 38
   source tracking, 40–41
Microflagellates, 112
Microplastics, 85
Millennium Ecosystem Assessment, 132
Minimata Bay, 69
Ministerial Conference on Environment and Health, 150
Modelling, 53, 146
Models for cholera, 55
Molecular
   -based methods, 38
   detection of
      microbes, 43
      microorganisms, 44
   methods, 42–43

Subject Index

Monitoring, 113, 116
  of shellfish toxins, 116
Mussels, 110–111

N

Nanomaterials within REACH, 84
Nanometre, 81
Nanoparticles, 12, 73, 80, 140
Nanoparticulates, 139
Nanopathology, 141
Nanostructures, 82
Nanotechnology, 84
Nanotoxicity, 141
Nanotoxicology, 83
Natural disasters, 137
Neurotoxic shellfish poisoning (NSP), 13, 104–105
Neurotoxins, 99
NORMAN research project, 74
North Atlantic Oscillation (NAO), 120

O

Organophosphorous compounds, 78
Over-exploitation, 129
Oysterbeds, 112

P

Paralytic shellfish poisoning (PSP), 13, 100, 109
Pathogen, 31–32, 53
  detection, 32
  distribution, 31
Pathogenic
  microbes, 31
  vibrio, 11
  viruses, 10
Pathogens, 2, 9–11, 17, 26–31, 36, 38, 53, 139
Pelagic microplankton, 113
Persistent Organic Pollutants (POPs), 73
Personal care products, 78–79
Pesticides, 75–76
Phaeocystis, 112

Pharmaceuticals, 78–79, 144
  from the Sea, 144
Phthalates, 76
Phytoplankton, 3, 13, 95–98, 100, 106, 114
  blooms, 97
  derived shellfish toxins, 100
  growth, 96
  monitoring, 114
  production, 97
    season, 96
  species, 106
Plastic debris, 85
Plastics, 12, 84–85
Point sources, 29
Policy needs, 15, 128, 146
Pollutants, 2, 9–11, 17, 29, 68–69, 73, 147
Polluted beaches, 39
Pollution control and waste management, 136
Polymerase chain reaction (PCR), 42
Polyomavirus, 40
Population
  decline, 136
  growth, 136
Precautionary
  approach, 140
  principle, 147
Priority
  pollutants, 11, 75
  substances, 71–72
Psychiatric pharmaceuticals, 79
Public
  health, 14
  perception, 70

Q

Quality of life, 134

R

Rapid Assessment of Marine Pollution (RAMP), 134, 141
REACH, 73, 84, 140, 147
Reactive oxygen species (ROS), 83

Red-tides, 98, 108
Regulatory policy, 133
Remote sensing, 55
Respiratory illness, 105
Risk, 137, 141, 143, 148, 150, 154
   assessment, 137, 141, 143
   management, 148
   reduction, 137

S
Satellite sensors, 55
Scallops, 110
Science-policy interface, 157
Seafood, 14, 26, 145
   -borne diseases, 26
   safety, 145
Sewage, 32–36, 42, 70, 77
   contamination, 33
   pollution, 42
   spill, 34, 35
   treatment
      facilities, 35
      plant discharges, 77
      system, 36
      works, 70
Shellfish
   aquaculture, 99
   beds, 113
   flesh toxicity testing, 114
   harvesting beds, 115
   Hygiene Directive, 114
   industry, 116
   poisoning, 99, 108, 117, 138
      events, 138
   safety, 109
   toxicity, 100, 118
   toxins, 109, 114
Silicoflagellates, 113
Stockholm Convention on Persisent Organic Pollutants, 147
Streptococci, 32–33
Sustainable industrial development, 136
Swimming in dilute sewage, 3

Systems
   approach, 16, 18
   -based approach, 151

T
Timescales of forecast, 21
Toxicity
   Identification and Evaluation (TIE), 86
   in scallops, 110
Toxins, 2
Tributyltin, 70

U
United Nations
   Convention on the Law of the Sea, 69
   Group of Experts, 69
   Industrial Development Organisation, 136
   International Strategy for Disaster Reduction, 138
   Stockholm Convention, 73
Urban Waste Water Directive, 15

V
Vibrios, 53
Viral gastroenteritis, 11
Virulence factors, 31
Viruses, 29

W
Water
   Framework Directive, 73, 143, 147
   quality criteria, 44
   Quality Framework Directive, 15
Wellbeing, 14, 106
World
   Health Organisation, 4, 57
   population, 6
   Tourist Organisation, 7

Z
Zooplankton, 3

Lightning Source UK Ltd.
Milton Keynes UK
UKOW030441150512

192582UK00001B/4/P